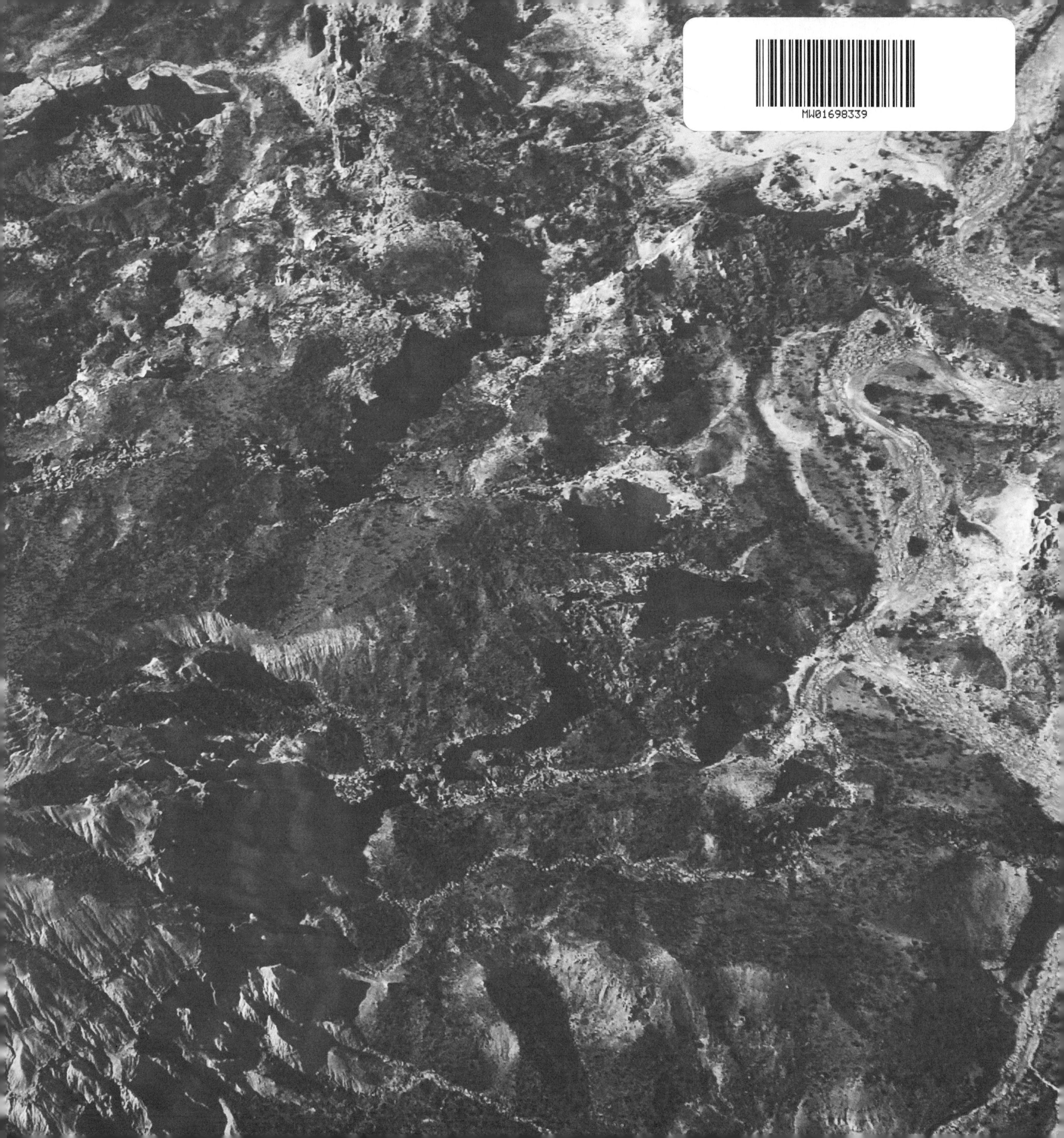

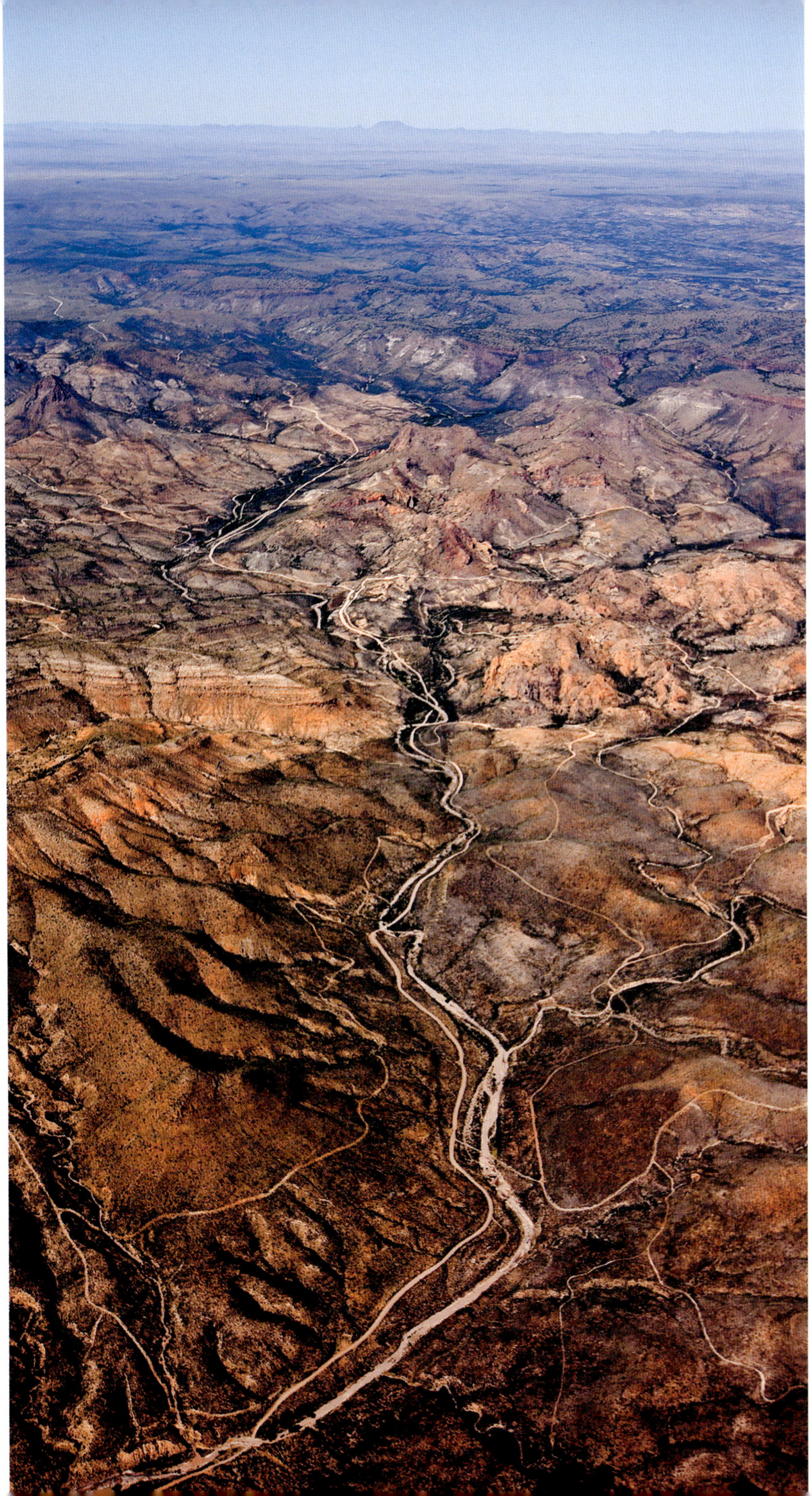

MARFA FLIGHTS

NUMBER TWENTY-SIX
Tarleton State University Southwestern Studies in the Humanities
T. Lindsay Baker, General Editor

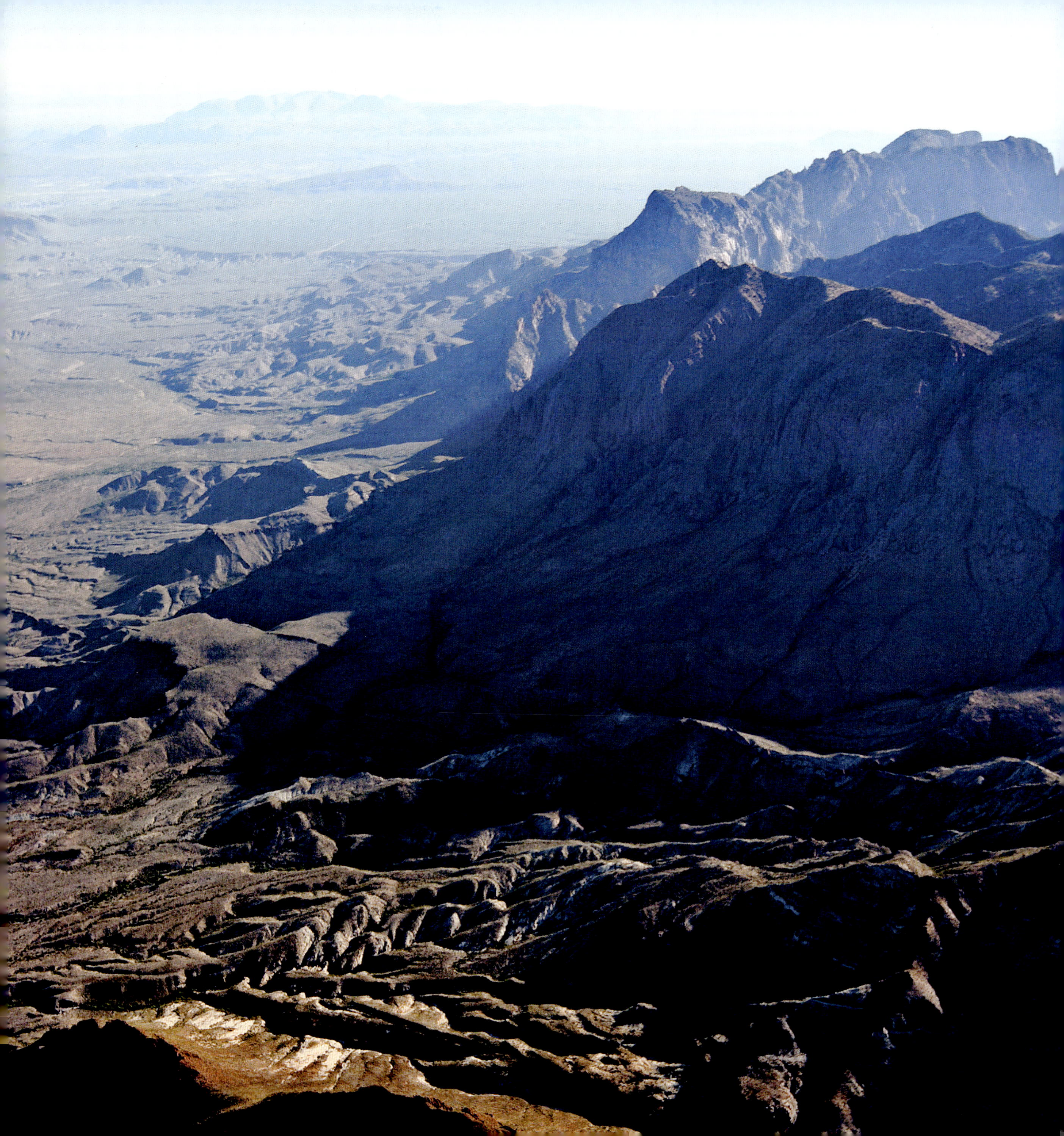

Marfa Flights

AERIAL VIEWS OF BIG BEND COUNTRY

Paul V. Chaplo

Foreword by T. Lindsay Baker

Introduction by Lawrence John Francell

TEXAS A&M UNIVERSITY PRESS COLLEGE STATION, TEXAS

First edition

This paper meets the requirements
of ANSI/NISO Z39.48–1992
(Permanence of Paper).
Binding materials have been
chosen for durability.
Manufactured in China.
∞

Library of Congress Cataloging-in-Publication Data

Chaplo, Paul V., author photographer
Marfa flights : aerial views of Big Bend Country / Paul V. Chaplo ; foreword by T. Lindsay Baker ; introduction by Lawrence John Francell. — First edition.
pages cm — (Tarleton State University southwestern studies in the humanities ; number twenty-six)
ISBN 978-1-62349-168-0 (flexbound : alk. paper) —
ISBN 978-1-62349-178-9 (ebook)
1. Big Bend Region (Tex.)—Pictorial works. 2. Big Bend National Park (Tex.)—Pictorial works. 3. Landscapes—Texas—Big Bend Region—Pictorial works. 4. Landscape photography—Texas—Big Bend Region. 5. Aerial photography—Texas—Big Bend Region. 6. Photography, Artistic. I. Title. II. Series: Tarleton State University southwestern studies in the humanities ; no. 26.
F392.B54C47 2014
976.4'93—dc23
2014000912

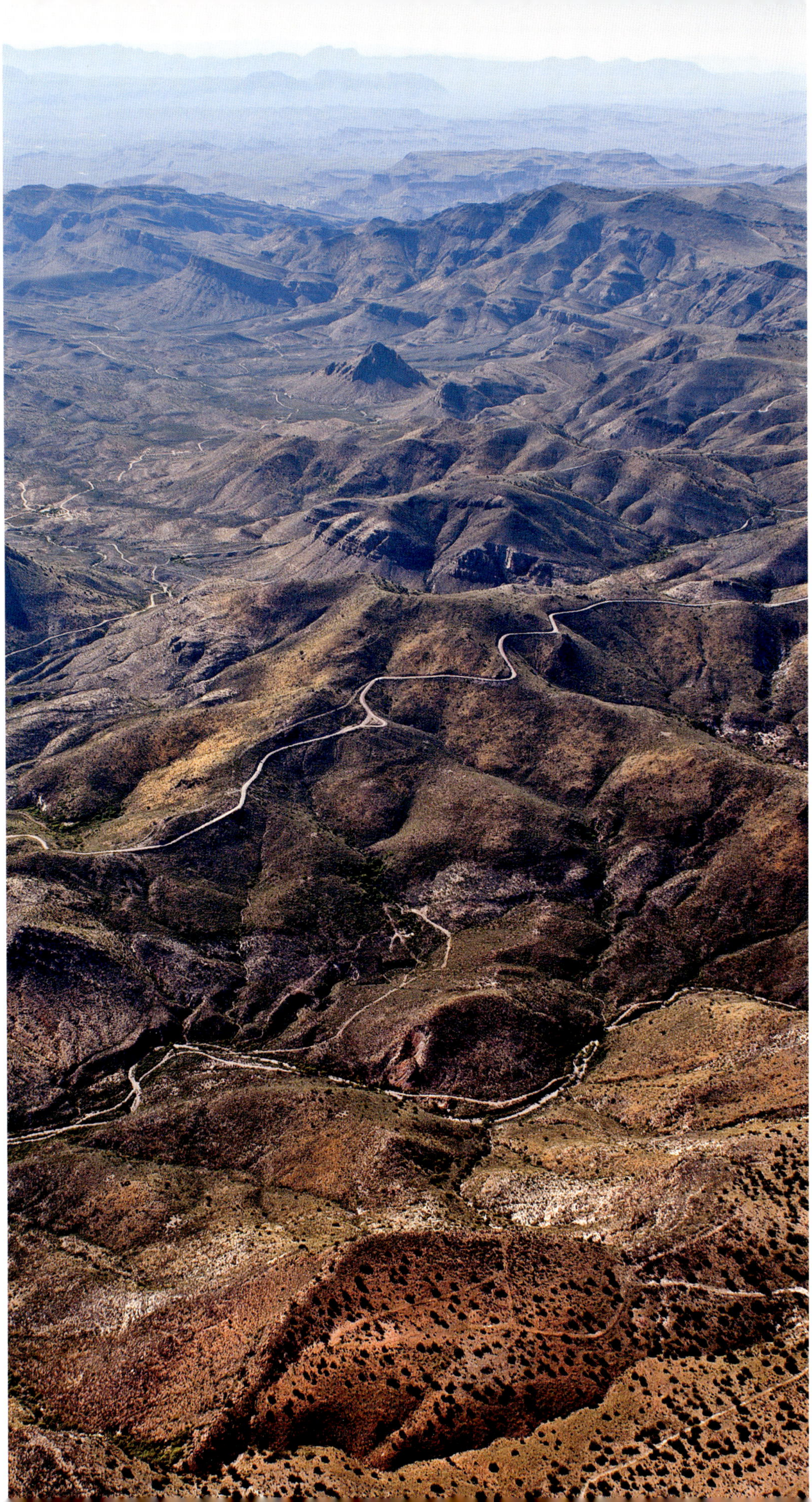

Dedicated in memory
of my father,
Maj. Victor J. Chaplo,
United States Air Force
combat and test pilot
and recipient of
the Air Medal
and the Distinguished
Flying Cross

Contents

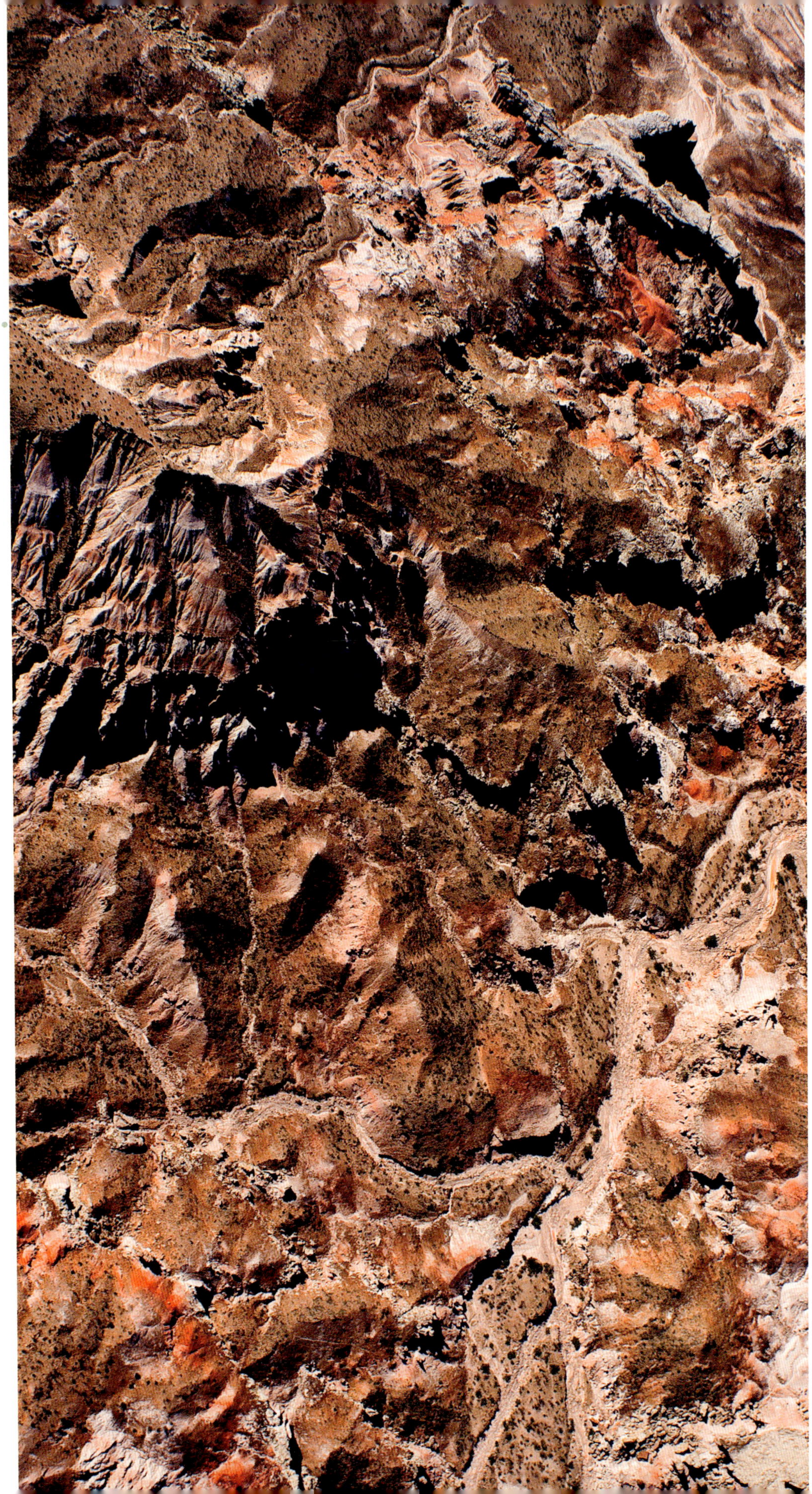

Foreword

Paul Chaplo and I first came together in 2003 at the W. K. Gordon Center for Industrial History of Texas in the Thurber ghost town, located west of Fort Worth. I had been hired to direct the new museum there for Tarleton State University, and the architects had selected Paul to make documentary photographs of the building they had designed. I observed at the time how particular he was about everything that had to do with this assignment. Paul later shared prints of his images with me, but in the rush of getting the new museum into full operation, I didn't think very much more about him for another three years.

In February 2006 Chaplo joined two other artists in putting together an exhibition for the gallery in the Wells Fine Arts Building at the Tarleton main campus in Stephenville. The show, *Texas, a Land Revealed,* included several of Paul's remarkable photographs made from the air above the Big Bend country. I was captivated by the combinations of landform textures making up his images. The show prompted the two of us to get back together. Eventually he even donated some black-and-white darkroom equipment for use at the Gordon Center museum.

Then in 2010 President F. Dominic Dottavio asked me to succeed Dr. Tom Pilkington as the editor for the Tarleton State University Southwestern Studies in the Humanities series at Texas A&M University Press. In this role I began taking the book series in fresh directions, dealing with the human impact on the land. As editor-in-chief Mary Lenn Dixon and I discussed potential new authors for the series, my thoughts returned to the photographer who had made the memorable aerial images of the Big Bend country. I described Paul's work to Mary Lenn, and she authorized me to explore publication possibilities with him.

I never expected to bump into Paul by accident in a commercial parking lot in Fort Worth, but that's what happened. To his great surprise, I asked Paul out of the blue whether he had ever thought about putting together a book of his Big Bend photographs. This was the last thing on his mind at that moment, but he agreed enthusiastically to consider the possibility. We later met to discuss the project in more detail at my office in Thurber, and there the concept for this book was born.

Since that time Paul and I have had many months of enjoyable and thoughtful collaboration. He has prepared truly superb images and combined them with an introduction and captions that provide context complementing his photographic efforts. These are efforts undertaken at times while hanging out the removed door of a Cessna 172 aircraft flying nearly sideways. I hope that the readers enjoy this book as much as I have enjoyed my role in connecting Paul Chaplo with the Texas A&M University Press.

—T. Lindsay Baker,
Tarleton State University

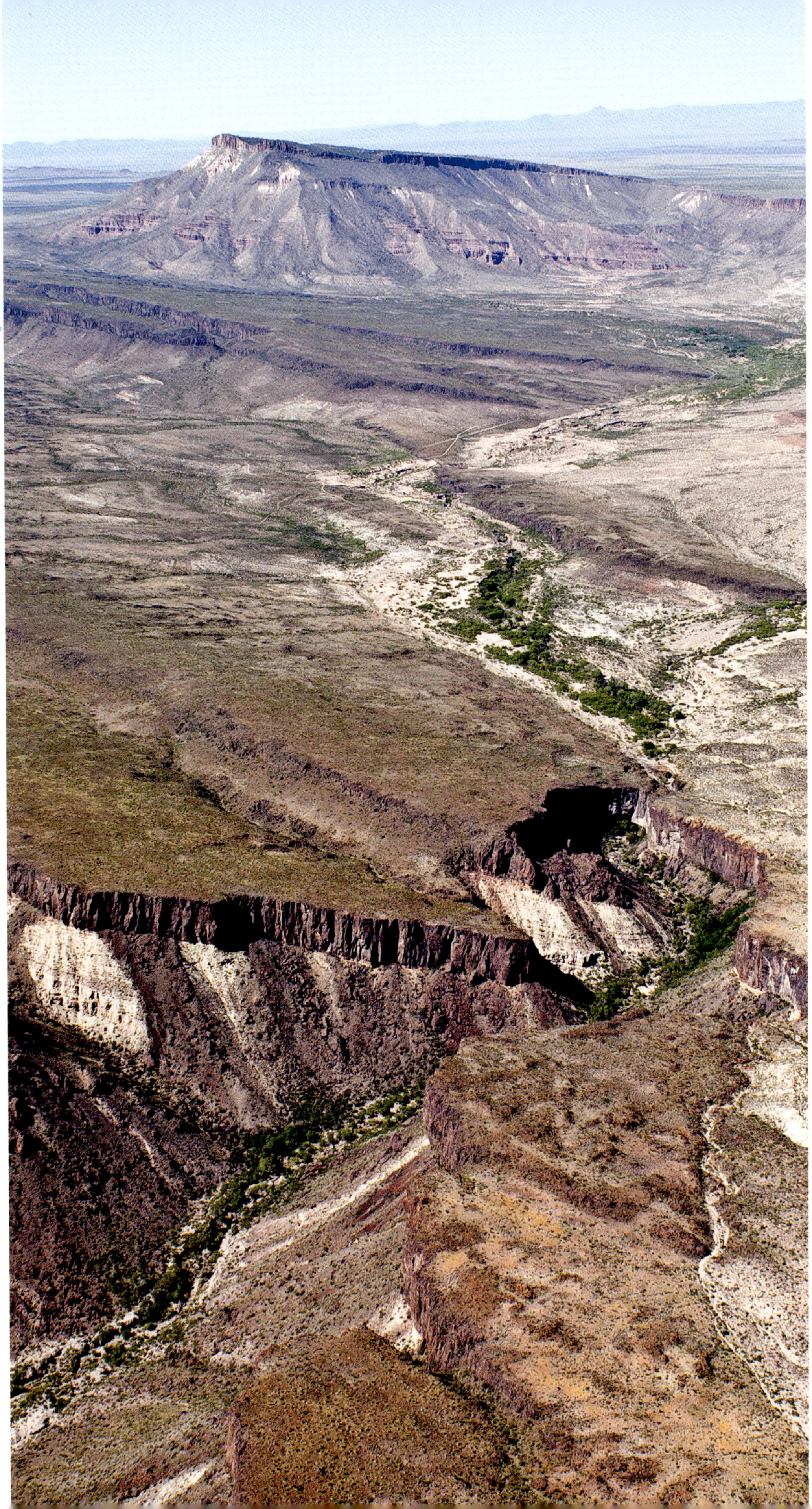

Photographer's Preface

I have a love for the land, which goes back to my childhood days of walking on my family's dairy farm and later hiking the Adirondack Mountains of upstate New York. I am a New Yorker who could live off the grid. When I came to Marfa, Texas, it was not for the art scene; it was for the austere beauty of the Big Bend country that surrounds it.

My interest in photography started at a young age when my grandfather showed me how to use a manual Exakta 35 mm film camera and a Weston light meter. The proportional thinking and creative options for camera, lenses, and exposure quickly became second nature to me. In addition, I learned traditional art skills, which have helped me to thoughtfully design my images. I suppose that my greatest influence as a photographer, strangely enough, is that I learned to draw. In learning to draw, I learned to design the space in the composition of an image.

My early art training was under New York abstract expressionists. I am always striving to balance formal composition of the image; technical concerns of camera, lens, and print; and the power of expressionist gesture. I still appreciate the work of artists such as Franz Kline and Robert Motherwell. Along with showing this land that is so meaningful to Texans, I hope that my images transcend "place" and open themselves to be experienced in a greater way as art. I continued growing as a photographer and eventually was awarded a fellowship in the Master of Fine Arts in Photography program at Rochester Institute of Technology in Rochester, New York. At R.I.T. I was greatly influenced by other committed photographers who were my classmates and world-renowned artists who visited and spoke with us personally about our work.

In terms of aerial photography, the image makers who most influenced me have been Marilyn Bridges (my classmate at R.I.T.), Yann Arthus-Bertrand, and Bradford Washburn. I particularly enjoy the strong black-and-white images of Washburn that combine sense of place, strong abstract composition, and awe-inspiring sense of scale. After graduating from R.I.T., I followed in Washburn's footsteps and photographed Denali and the glaciers of Alaska, flying with K2 Aviation. I loved the drama of mountains, snow, and ice. Like Washburn, I was attracted to both the vistas of the mountains and the details evident in close-ups such as the texture of a glacier, showing myriad crevasses or a pool of emerald water on the ice.

Resource-rich Rochester is a wonderful place for a budding photographer. Like Minor White, I learned Zen meditation—in my case at Philip Kapleau's Rochester Zen Center. Trips to the International Museum of Photography at the George Eastman House exposed me to the full array of work made by those who came before me. In addition to my studies at R.I.T., I attended the Visual Studies Workshop where William E. Parker helped me to see the archetypal power in personal imagery in his course on Jungian visual analysis.

I suppose that if I were a musician, one might say that I was classically trained. At R.I.T. in the master's program headed by Dr. Richard Zakia and then Ken White, I learned the Ansel Adams "Zone System," then explored the more mystical approach of artists such as Minor White that seemed kindred to my innate preferences. The strong technical orientation of R.I.T. instilled me with a willingness to use new tools, which made my transition to digital photography seamless and even enjoyable.

Despite all the art training and exposure to wonderful exhibits in New York museums and galleries, some of the most striking images that have impacted me go back to my childhood, being raised in a US Air Force family. My father was a combat pilot and had a collection of books on the history of aviation that I spent many hours studying. Always in my visual memory are haunting images that I have seen made through military gunsight, bombsight, and reconnaissance cameras that show the scars of war on the earth.

These photographs, made from airplanes, balloons, kites, and even pigeons, give a strangely detached view of life-changing events. I feel the same about Google Earth imagery and aerial mapping—while showing how the earth is changing and allowing us to navigate conveniently, such images to me are the opposite of my goal. I strive to create evocative compositions that allow the viewer to see the place in a new way while leaving room for personal interpretation and reflection.

In terms of process, I work fast and take lots of pictures. The aircraft is always moving and the view and perspective are constantly changing. I enjoy the dynamics of working from the air and collaborating with a skilled pilot. In photographing Big Bend country, I was particularly drawn to views that showed the rectilinear marks of mankind engulfed by the massive natural forms of the landscape.

There were some memorable moments in photographing Big Bend that may reveal what catches my eye and why I chose certain images. The first is *End of the Road* (8). I saw the possibilities for this shot and asked my pilot to turn to allow a better camera angle. The immediate appeal to me was the juxtaposition of man-made and natural elements. Ruler-straight Pinto Canyon Road cuts a line across the flatlands, heading for the massive barrier of the Chinati Mountains, where the diminutive road will be engulfed by nature. This to me speaks of the ability of Big Bend to reassert nature's dominance over man's effort to subdue it—seemingly to be a last stand against civilization's marks on the landscape.

Another image that comes to mind is *Langford Hot Springs* (50). Here a well-known destination for Big Bend visitors, Langford Hot Springs, is shown in a vertical aerial photograph, looking straight down on the landmark. The tin-roofed building, now tiny, is shown in the awe-inspiring context of the Rio Grande and the surrounding hills. The terrain is textured and naturally terraced by erosion. To me, it looks like a USGS topographic map come to life in the landscape, with the contour lines physically scribed onto the mountains to show elevation. I like the way that the intersection of the Rio Grande and Tornillo Creek divides the land into massive triangular shapes.

Here the design prevails, not as triviality but as underlying significant structure, and a familiar spot is presented new. I chose this image for those reasons and for its place in the sequences of photographs that precede and follow it. My training in cinematography and work as a professional video editor comes into play in how I select and order the work in a book or exhibit. In this case, to me, the sequencing is similar to a film-editing cut to a close-up after a wider establishing shot.

In this way, there is a visual narrative that is determined by creative decisions in each photograph and by the ordering of images in a sequence. Underlying all of this is the role of the aircraft as part of the mise en scène—it is the ultimate dolly and crane. Then, on a personal level, there are my impressions and openness to what significance and meanings the images have for me, such as emptiness, openness, desolation, refreshment, euphoria, foreboding, or freedom. Then I consider how I experience meaning in the relationship between images that appear adjacent to each other, either physically on the wall or page, and even stacked, shuffled, or blended in an individual's memory.

To make a living, I work as an independent commercial photographer and aerial video producer based near Dallas, Texas. Most of my production time is spent capturing architectural, energy, and infrastructure projects from a helicopter. Working in the congested Dallas airspace where aircraft abound, often maneuvering at very low altitude around obstructions, has forced me to sharpen my aerial photography skills. I sometimes feel that I face death when I go to work.

One morning, I clearly remember flying back from an assignment southwest of Fort Worth. The return flight took our helicopter along the shore of White Rock Lake just northeast of downtown Dallas. As we flew along the sailing club, with its numerous docks and boats along the S-shaped shoreline, I snapped a photograph that seemed to perfectly balance the water and land. Then it struck me that I could take my hard-earned skills, at a time when I was at the top of my game, and apply them to my passion—the landscape of Texas.

To me, this was a heartfelt yearning, as my hard work to survive in the photography business had seemingly taken me away from the thread of my soul that is my artistic life. I had wondered how I could even reach that thread and connect again. Suddenly, in an epiphany, I saw the chance to make a running start and take a leap of faith to reach it.

Immediately I thought of Big Bend as the dramatic and uniquely Texas landscape that would make the ideal subject for my artistic aerial photography. My sojourn to Marfa could be a means of reconnecting with my love for art and the land. There were many challenges, including saving money for the project, finding an airplane, scheduling the flights, and, most importantly, locating the right pilot. Roger Amis appeared on the scene as a world-class aviator and much more—he knew every nook and cranny of this vast country and became my guide for this adventure.

Once I arrived in Marfa and took to the air, the experience was more than I expected or could comprehend. I saw views that most people will never see in their lifetime. The beauty of the place bowled me over, and all the time I was making photographs to bring back and share with my family and friends. I was immersed in the artistic process and my creativity was flowing "full-bore."

I remember that first morning very clearly. It seems like it was just yesterday . . .

Before daybreak on the morning of the first flight, I meet my pilot for breakfast at a hole-in-the-wall cafe in Marfa where they serve scrambled eggs with tortillas and homemade chipotle sauce instead of toast and jam. He is smiling and good-natured—and it seems that everybody knows him. His name is Roger, and he is a former army attack helicopter pilot and recently retired US Border Patrol Aviation pilot in Marfa. It has taken over a month to convince him to fly for me on a series of aerial photography missions over the Big Bend country to the very edge of Mexican airspace.

The coffee is strong and I need it because I've only slept a few hours in the affordable motel at the edge of town. I spent most of the night cleaning my lenses and checking my cameras. I have to remind myself not to drink too much coffee because once we take off there will be no toilet.

Arriving in the dark at the diminutive Marfa airport, I see the Cessna 172 waiting on the tarmac. Roger shows me the flight plan in the small flight operations office and tells me that Ronnie, the ground crewman, is making final checks on the plane. My pilot "called in" a favor and borrowed the Cessna from a friend in another city and flew it to Marfa because there wasn't a suitable aircraft available locally. The charter costs more than I pay for my corporate photography flights around Dallas but there's no alternative.

Dressed in jackets to ward off the desert cold, we walk to the plane where I meet Ronnie for the first time. He is obsessively checking the aircraft. There is a bright yellow kitchen sponge duct-taped to the outside of the aircraft window on the side where I will sit. We joke about its looking like cartoon character SpongeBob. The hinge restraint has been removed from the window to allow the aerodynamic lift from the wind to hold it open against the underside of the wing so I can take unobstructed photographs. The sponge will keep it from gouging the wing during flight.

Ronnie shows me the inventive sponge arrangement as he applies another strip of silvery duct tape to the Plexiglas window. He says, "Roger usually doesn't take people where you want to go, but he's taking you. He ran a background check and says 'you are who you say you are.'" In the meantime, my pilot is loading desert survival gear into the rear compartment. The in-depth preparations demonstrate the risk of flying over remote areas of Big Bend. I feel grateful for Roger's Border Patrol training. He has stowed away sleeping bags, tent, water, trauma kit, and an emergency radio.

I load my camera bags into the Cessna as Roger checks the fuel level again and prepares for takeoff. He smiles when he sees the rescue strobe attached to my camera bag. On his jacket he has clipped a serrated knife for cutting seatbelts so that neither of us becomes trapped in wreckage in the event of a crash landing.

We climb aboard. The pilot opens his window and yells, "Clear prop!" The motor shudders and the prop jerks, then spins smoothly as the engine starts. He asks me, "Ready?" I give him the thumbs-up, and we taxi to the runway. "You are about to see some things that few people alive have ever seen," Roger says. The aircraft accelerates down the runway and lifts into the cold desert air. I take my first photograph:

the shadow of the aircraft racing across the desert floor a hundred feet below. The desert transitions to rolling hills that look like soft undulating velvet. At the edge of the hills runs a lone road, first straight like a ruled line across the desert, then winding serpentine as it encounters the change in terrain. As we climb and Marfa fades into the desert horizon behind us, the marks of humans become smaller and smaller within the expansive, austere landscape.

The desert plateau suddenly drops away below us and we are over a canyon. We work at low altitude. Roger informs me that if we lose power now, there is no place to land in an emergency. Then we climb higher and higher to prepare for a vertical aerial photograph looking straight down on the canyon. I secure my camera bag and open the window. Roger puts the plane into such a tight turn that the craft flies on its left side.

Outside my open window, straight down, the canyon has replaced the horizon and presents a mesmerizing view, spinning centered in my camera viewfinder. The safety harness keeps me inside the plane as I fight the g-forces of the tight turn to hold the camera steady as I frame the shot to capture the image.

The plane is rapidly losing altitude, so I have to work quickly before Roger terminates the maneuver for safety. I can hear the airspeed increase as we fall, spinning toward the earth as I keep shooting. I see the view I've been hoping for: the curving river and eroded land forming a visual balance like a ying-yang symbol made of about twenty-four square miles of land. We are within a hundred yards of Mexican airspace.

After working along the Rio Grande and the Chisos Mountains, we notice that fuel is running low. We turn north to Marfa. As we head home, dust devils spin below us as the sun, now higher in the sky, heats the Chihuahuan Desert. Roger lets go of the steering yoke and points to the rising altimeter reading. The plane is being lifted on the desert thermals, soaring like my spirits. Flying home, I feel that everything in my life has led to this, and I was prepared for this moment. In a way it's as though I was in training for this highlight of my life, what I live for, my bucket list item. To me this was the ultimate—to fly over Big Bend and make fine art photographs.

I photograph museums as part of my professional architectural photography career. One such assignment took me to the W. K. Gordon Center for Industrial History of Texas in Thurber, where I met Dr. T. Lindsay Baker, who is the director there. I knew of Baker from his many books on the history of Texas, and I was especially familiar with his popular *Ghost Towns of Texas* series. When I had completed my assignment at the museum, I shared some prints with him.

All the while, I continued working on the Big Bend images and loved the look and feel of the prints. As Minor White might say, the spirit came down for this project. I had a couple of regional exhibitions including a group show at Tarleton State University. I was pleasantly surprised to see Dr. Baker at the opening of the event.

Some time passed, and then while I was on an errand for photo supplies, we ran into each other again. Dr. Baker suggested the possibility of publishing the Big Bend work. Later we set up a meeting, and I drove to the Gordon Center. It was a hot day, about 103 degrees F, and while driving west on the big interstate highway, the air conditioning in my SUV failed. I stopped at a gas station and tried to fix it to no avail, so I bought a tall iced tea, got back on the road, windows open, and continued on my way.

I remember, for one of our meetings, walking into the museum in Thurber with a pile of mounted 17-by-21-inch prints. I had to make a couple of trips to my vehicle to unload all of them. We put the large photographs around the inside perimeter of Baker's office like an improvised exhibit. T. Lindsay immediately identified places such as Boquillas, where he pointed out details that I missed, such as the remains of the old cotton fields, and told me about his crossing of the Rio Grande on the back of a burro. He generously shared his knowledge of the publishing process, and I soon developed a prospectus.

Working closely with Texas A&M University Press, Dr. Baker then preselected my manuscript for inclusion in the funded series Tarleton State University Southwestern Studies in the Humanities, which conferred many advantages. The greatest of these has been his mentorship throughout the process.

Acknowledgments

Special thanks to the stellar team at Texas A&M University Press for their creativity and professionalism in refining and publishing this book. This book was realized with the essential love and encouragement from my wife Cyndi, a true renaissance woman who combines many talents as musician, artist, and engineer. Also, a big Texan "thank-you" to Tarleton State University for its support of this work as part of its Southwestern Studies in the Humanities series. Dr. T. Lindsay Baker, the series editor, provided legendary support as my mentor and guide while I navigated the winding path of academic publishing for the first time.

With gratitude, I acknowledge the following people for their support in the creation of this book and the photographs: Lawrence John Francell, John Morlock, Roger and Jeanne Amis, Priscilla Chaplo, Anne Marie Chaplo, Michael Chaplo, Chris and Julie Roberson, Grant Carpenter, and Ron Lewis.

As we begin to schedule exhibitions of the fine art photographs seen in this book, I express sincere appreciation to Elizabeth Jackson and Mary Bones at the Museum of the Big Bend in Alpine, the first major venue to welcome the show.

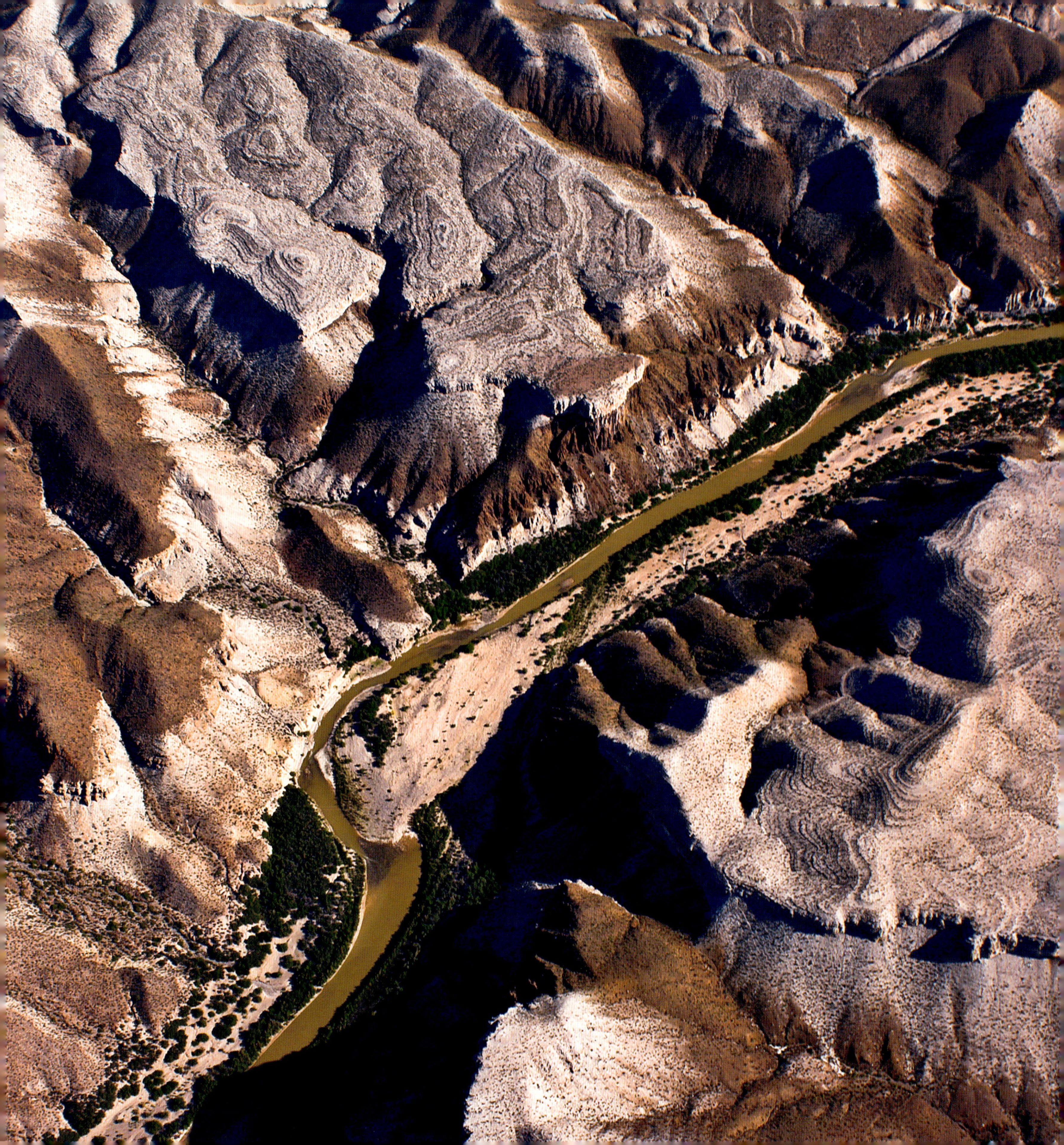

Introduction

Lawrence John Francell

The Texas Department of Public Safety helicopter took off from the Alpine airport with four of us on board. It was several weeks after the devastating Rock House Fire in April 2011. The pilot and copilot were DPS officers and I was a passenger with Brian Powell, who managed the area office for state representative Pete Gallego. Representative Gallego was in Austin and had Brian survey the fire damage. In my capacity as commissioner for Jeff Davis County, I was asked along.

I had never flown over our country at low altitude and was excited to finally have the opportunity. At first the experience was disconcerting. As a historian and map collector, I am used to knowing where I am in reference to the land, but from the air I quickly lost all orientation. Without the compass and GPS in the control panel, I was lost. And then it did not matter anymore. Even with the land black from the fire, the view was magnificent, and north, south, east, or west was no longer of concern. The beauty of the landforms, mountains, rock palisades, dry river courses, and an occasional green oasis where springs leaked from the bluffs leaves one with a sense of awe. This is the same feeling one has when looking at the photographs of Paul Chaplo.

On the ground in the Big Bend country everything seems to either stab, stick, sting, or bite, but from the air the vegetation merges into the rocks and the rocks into landforms to create a seamless beauty, providing images of artistic as well as documentary interest. The viewer is swept up into a way of looking at the land that is unique and where place is less important than form.

To see the environment from above has long intrigued mankind. One of the most successful art forms of the nineteenth century was the Bird's Eye View lithographic print of one's community. No self-respecting town would pass on the opportunity to commission and promote such an image. A number of artists and print companies of the day became both successful and famous for their work. The development of inexpensive lithographic printing methods that could successfully reproduce multiples of an exact image flourished in this period, becoming what Peter Marzio called the "democratic art." Artists fanned out over the countryside to produce these town and city views for commercial sale at reasonable prices that any self-respecting citizen could afford. According to Dr. Ron Tyler, retired director of the Amon Carter Museum, there are over seventy of these views of Texas towns. Over 2,400 towns and cities in America are documented in this way.

The problem with the bird's-eye view is that each artist had a different style and, as a commercial product, each image has to show the community in the best light. The sun is always shining, industry and railroads yards are clean, and the grass and trees are green. As Gerald Danzer, an expert in this field, has written, "It might be best therefore, to think of these views as flattering urban portraits that follow certain rules of perspective, setting and mood, and overall style. Because they were sold as commercial ventures to the townspeople, they had to look accurate to be convincing, but not so honest as to reveal the problems and imperfections of their subjects." Thus the bird's-eye view is interesting as art or illustration, but ultimately unsatisfying as an accurate historic document establishing an aerial perspective.

With the development of photography, the world of art and the ability to document what was perceived as reality would change. In 1839 the Frenchman Louis Jacques Mande Daguerre announced his photographic process to the world. Beginning with the Daguerreotype, new and improved processes were developed so that by mid-century photography was a common medium. In 1858 another Frenchman, who called himself Nadar, took to the air in a balloon and created the first aerial photograph. It was in Boston in 1860 that James Wallace Black made the first aerial image in this country. His photograph of downtown Boston, made at a height of 1,200 feet, was described by another local photographer, Oliver Wendell Holmes: "Boston, as the eagle and wild goose see it, is a very different object from the same place as the solid citizen looks up at its eaves and chimneys." The need to reach for the sky in order to look back upon the earth and document what one saw seems to exist with the same passion as photography itself.

While there are photographic images from the US-Mexican War and the California gold rush, the public image of the West was still represented through paintings and prints in the work of such artists as Alfred Jacob Miller, George Catlin, and Karl Bodmer as well as the numerous official government publications such as the *Pacific Railway Surveys,* Joseph Christmas Ives's *Report upon the Colorado River of the West*, and William Emory's *Report on the United States and Mexican Boundary Survey.* Those, plus numerous others, received wide distribution to a country curious about the West.

It was the Civil War that established photography as the great documentary medium. While tethered observation balloons were used during the war, there seem to be no aerial photographs that exist. However, photographs seem to be present for virtually every battlefield as well as numerous participants, great and small.

After the Civil War photographers and artists returned to the West to create work that expressed their own reality. Whether it is Timothy O' Sullivan's work with the government surveys of Lt. George Wheeler and Clarence King; Andrew Russell documenting the construction of the Union Pacific Railroad; or William Henry Jackson, whom William Goetzmann calls the "greatest of all Western photographers," the photographic record of the last decades of the nineteenth century are magnificent in every respect. Yet at no time did any of these men rise above the height of the rim of the Grand Canyon or a peak in the Rocky Mountains to look down from above.

The world of photography would change once again on a day in December 1903 when the Wright brothers made the first manned flight at Kitty Hawk, North Carolina. No longer would the camera be earthbound. Antoine de Saint-Exupery, the French aviator and author of *The Little Prince*, wrote, "The airplane has revealed for us the true face of the earth. . . . We set our course for distant destinations. And then, only, from the height of our rectilinear trajectories, do we discover the essential foundation, the fundament of rock and sand and salt in which here and there and from time to time life like a little moss in the crevices of ruins has risked precarious existence." He could as well be describing Paul Chaplo's Big Bend country.

The airplane established a stable platform for a camera as well as a platform where the speed, height, and direction could be controlled. In the early years of aviation aerial photography was mostly a novelty, but it became serious business and an aid to victory during World War I. While the conflict raged in Europe, life along the border between Mexico and the Big Bend country of West Texas saw its share of conflict that eventually involved military aircraft and aerial photography.

The Mexican Revolution began in 1910 as a response to the oppressive policies of President Porfiro Diaz and a desire by many for land reform. The revolution was led in northern Mexico by Pancho Villa. There was as well a general lawlessness that greatly disrupted life along the border. Villa's attacks on the town of Ojinaga and El Paso and border raids by bandits at Glenn Springs in the Big Bend and the Brite and Nevill ranches farther up the Rio Grande brought the US Army and military aviation to the border.

In June 1919 the US Army Border Air Patrol was established, with initial airfields at Fort Bliss (El Paso), Del Rio, and Marfa. Lt. Stacy Hinkle, one of the original pilots of the service, described those early days, "Since flying was then in its infancy, knowledge was scant and equipment was primitive and unreliable. The established landing fields were few, and the servicing of planes and motors away from them was nil. Cross-country flying over any extended distance for groups of planes was at that time practically unknown." Equipped with British-designed and American-produced DeHaviland two-seat DH-4s, the Border Service took to the air.

The vagaries and danger of patrolling over the desolate and unfamiliar Big Bend is illustrated by the adventures of Lt. G. H. Peterson, pilot, and Lt. P. H. Davis, observer. On August 10, 1919, while flying on a routine patrol from the Marfa Airdrome to the Rio Grande at Lajitas and then up the river toward El Paso, they became lost, possibly mistaking the Rio Conchos for the Rio Grande. Eventually running out of fuel, they were forced to land. On August 17 Jesus Renteria, one of the region's insurgents associated with Pancho Villa, turned up at the store owned by Dawkins Kilpatrick at the border community of Candelaria. He demanded $15,000 for the return of the two flyers and wanted the funds by the next day.

Unable to acquire the ransom money through military channels, Col. George Langhorne, Eighth Cavalry, commanding at Marfa, approached a number of local ranchers and businessmen for assistance. At the time, much of the population of the area was in attendance at the Bloys Camp Meeting, an annual interdenominational religious event at Skillman Grove near Fort Davis. The news arrived in the afternoon during the daily men's prayer meeting and within minutes the ransom money was subscribed.

The task of delivering the money was given to Capt. Leonard Matlack, commanding officer of Company K, Eighth Cavalry. Known as a tough professional soldier, Matlack crossed the river at midnight on August 19 to rescue the American flyers. Paying half of the ransom, he returned with Lieutenant Peterson. Crossing the river once more, he was able to rescue Lieutenant Davis and escape without paying the other $7,500. Immediately three hundred troops of the Eighth Cavalry crossed into Mexico in pursuit of the bandits. In the lead was a Border Service aircraft.

The experiences of Peterson and Davis illustrated the need for better information and, especially, aerial maps of the region. These were soon produced as military aircraft mapped the entire border from Brownsville, Texas, to San Diego, California. These current photographs by Paul Chaplo follow in a long tradition of seeking to understand and find our place in one of the last frontiers of America.

It was not only the army that developed an interest in documenting the Big Bend from the air. The first civilian to take such an interest was Wilfred Dudley Smithers. While the army photographers were primarily interested in charting the ground as an exercise in self-preservation, Smithers went much farther afield. His was a fascination for the land itself: "The Big Bend of Texas is a world unto itself, a rugged forbidding peninsula of land left remote and isolated by the irresponsible antics of the Rio Grande. This geologic fantasy land is the perverse creation of that mighty river as it plunges deep into Mexico and then, as if changing its mind, partially reverses its course in its search for the distant waters of the Gulf of Mexico." It was not until Paul Chaplo came along that anyone documented the Big Bend from the air with such vision.

W. D. Smithers was born in San Luis Potosi, Mexico, on August 31, 1895. His father was the bookkeeper for the American Mining and Smelting Company. Smithers would always maintain a connection to Mexico, later documenting the land and peoples of the northern part of the county. In 1905 the family moved to San Antonio, where he learned photography at a local studio. He did not finish high school, but by 1912 he owned his own camera. From 1915 to 1917, during the border troubles, he worked as an army teamster using pack mules throughout the Big Bend to supply the soldiers. Included in his photographs from these days as a mule skinner are scout planes being used in support of ground operations.

In 1917 Smithers enlisted in the army, first in the cavalry but transferring to the Aviation Section of the US Army Signal Corps. There he developed a passion for flying and was able to experiment with aerial photography. The problem faced by the aerial photographers in the day of aircraft with open cockpits was the air current produced by propeller backwash. Early cameras with flexible rail-mounted bellows to adjust focus were not adequate. Smithers built his first camera when he was thirteen and had already faced the problems of durability and portability while packing mules through the Big Bend. He quickly found a solution to the problem: "To withstand the high winds during aerial photography, I used a wooden cone rather than a bellows." Over the course of his career this harsh country would dictate that he would have to build much of his equipment and improvise ways to develop and print his work in the field.

From 1920 until 1929 Smithers operated a photographic studio in San Antonio and worked freelance for the local newspapers. He continued his interest in aerial photography, making numerous trips to the Big Bend and Mexico. His connection to the army allowed him the unofficial use of military aircraft, often seeing his images printed with an "official" Army Air Corps imprint. In 1933–34 he documented the development of the Davis Mountains State Park Highway at a time when the Civilian Conservation Corps was creating Davis Mountains State Park and the University of Texas was building McDonald Observatory. In 1935 he would document the US-Mexican border for the Immigration Service, and, as momentum gathered to create Big Bend National Park, he produced a series of photographs to promote this effort.

For this documentary work Smithers would produce lantern slides, a favored method for illustrating public lectures and travelogues of exotic places. These glass slides, 3¼ by 4 inches, were enhanced by hand-tinting the black-and-white images. Over time he produced over 9,000 images. In his 1976 autobiography, *Chronicles of the Big Bend,* he wrote, "My photography should direct itself to historic and transient subjects—vanishing lifestyles, primitive cultures, old faces and odd, unconventional professions. Before my camera I wanted huts, vendors, natural majesties, clothing, tools, children, old people, the ways of the border." As well, he was the first to capture the natural majesties of the Big Bend from the air.

There is no finer description of the natural wonder that is this great bend in the river than that of Ross Maxwell, geologist and first superintendent of Big Bend National Park, who wrote, "According to Indian legend, when the Great Creator made the earth and finished placing the stars in the sky, the birds in the air, and the fish in the sea, there was a large pile of stony material left over. Finished with His job, He threw this into one heap and made the Big Bend." As man discovered the region and succeeding groups of people encountered this large pile of leftover stony material, they each had to find ways to survive and even thrive in this environment, but none ever tamed it.

That prehistoric Native Americans roamed far West Texas is known and intensely studied by the archeologists working at the Center for Big Bend Studies at Sul Ross State University. By the time the Spaniard Cabeza de Vaca entered the region in 1535, the native peoples were settled in agricultural communities along the Rio Grande, especially at La Junta, the junction of the Rio Grande and Rio Conchos near present-day Presidio.

Named Jumano by Antonio de Espejo during his entrada of 1582–83, this culture actually consisted of two groups with different lifeways. One group lived in sedentary Puebloan-style communities, farming the alluvial banks and fans of the river. The other group, nomadic hunters, roamed the Big Bend and mountains of West Texas out to the southern plains. Both faced the day-to-day issues of shelter and protection, each in different ways. Along the river they gathered in groups, but the nomadic peoples found protection in seeking the high ground, the mesa tops where they could keep watch. It must be in the nature of man to view the world from a height as a way to explore and understand his surroundings.

Tradition tells us that the Spanish pushed north from the interior of Mexico for God, glory, and gold, but sadly for them not much of the latter was found. At one time missions were scattered up and down the Rio Grande, but most of them eventually failed as there were no riches to be found, and the only glory was in survival. What the Spanish did leave was a rich culture of language in the names of places and landmarks.

The people who would survive the best, at least until the arrival of the Americans, were the nomadic tribes with a warrior culture, the Apache and Comanche. With a lifestyle based upon periodic raids interspersed with trade with the Spanish and Indian settlements, these tribes viewed what many saw as a desolate environment as one rich in resources. This land was a true home, and they would fight long and intensely to protect it, and would continue to do so until met by overwhelming force in the form of the US Army.

The first Americans to arrive were a combination of adventurers and soldiers surveying roads and determining boundaries. Ben Leaton, by legend, arrived in the region as a member of Alexander Doniphan's column marching south from Santa Fe during the US-Mexican War. In August 1848 Leaton became the first American to own land in the Big Bend, creating his own fort and private trading post east of Presidio. While Ben Leaton saw the Big Bend as a place to exploit commercially, his neighbor Milton Faver saw the potential for livestock.

Faver arrived in the area around 1858; the actual date is not known. He started the first cattle operation in the region, controlling a vast region by owning the water sources necessary for his cattle. With an abundance of water Faver was able to raise fruit trees, and he is most remembered for his peach brandy. With no transportation or anything resembling a reasonable market, turning peaches into brandy and using that as trade merchandise made business sense. As pioneers, both Leaton and Faver left their marks on the land, one with a state park and the other with Cibola Creek Ranch, a resort at the restored El Fortin del Cibolo.

The end of the war with Mexico and the discovery of gold in California created an immediate need to develop a transportation and communication network as well as the need to determine the precise boundaries of the new territories. The instrument of these endeavors was the army, especially the Corps of Topographic Engineers. Their activities would bring the first systematic and scientific exploration to the region. With the discovery of gold in California the officer commanding military forces in Texas, Gen. William Worth, understood the need for a road from San Antonio to El Paso. He assigned Lt. William H. C. Whiting and Lt. William F. Smith to the task. On the outbound trip they passed through the Davis Mountains and the Big Bend country to Presidio, hoping to find enough water for an acceptable road. Water was not to be had and the land too rugged, so on the return trip they surveyed a trail that looped south of the Davis Mountains that would soon provide the major all-weather road to the gold fields.

A more thorough and scientific exploration of the Big Bend was that of the United States and Mexican Boundary Survey. A complex and difficult task, made more complicated by politics and the lack of correct maps, the survey, authorized in 1848, was finally brought to completion by Maj. William H. Emory. His final report was presented to Congress in August 1856. His survey through the Big Bend provides for the first time a description of the country by a trained engineer and observer. "No description can give an idea of the grandeur of the scenery through these mountains. There is no verdure to soften the bare and rugged view; no overhanging of trees or green bushes to vary the scene from one of perfect desolation. Rocks are here piled one upon another, over which it was with the greatest labor that we could work our way," he writes. The Big Bend proved to be the most difficult terrain crossed by the boundary survey. As one studies the photographs of Paul Chaplo, these difficulties become clear.

Emory, Whiting, and Smith saw the land thought the eyes of engineers and soldiers on specific missions, but they left their marks on the land, especially Emory. He describes how the foremost landmark in the Big Bend was named: "Whenever the spectator was elevated sufficiently to see beyond the valley of the river, two prominent peaks were always present to his view: one of these marks a summit in the range of the Mexican Sierra Carmel; and the other, from its peculiar shape and great height, was long and anxiously watched during the progress of our survey." Several days would pass before the expedition approached this peak, which turned out to be one of a cluster of mountains called Los Chisos. In his report he writes, "For this peak we have proposed the name Mount Emory." Besides Emory Peak, the focal point of the Chisos Basin, there is also the Emory oak (*Quercus emoryi*) and the Emorybush (*Emorya suaveolens*). Each subsequent group that called these mountains home or passed through on important business began to fill in the names as if that alone could tame this country.

In the 1880s the railroads came and the Indians went. In January 1883 Collis Huntington's Southern Pacific completed a transcontinental line across the state, beating Jay Gould's Texas & Pacific Railway to the punch. The Southern Pacific crossed north of the Big Bend and south of the Davis Mountains on its way between El Paso and San Antonio. Gould's Texas & Pacific, caught short in an attempt to reach El Paso from Fort Worth, had to join tracks with the Southern Pacific at Sierra Blanca. These railroads, especially the Southern Pacific, would bring major changes to the region primarily through the creation of towns at regular intervals. Valentine, Marfa, Alpine, Marathon, and Sanderson, creating a divide between the mountains, also created gateways to the Big Bend.

Virgin lands, albeit lacking water in many places, and the coming of the railroads created a minor land rush as cattle and sheepherders sought grazing land. Seeing the unspoiled grasslands as the raw material for their livestock factories, ranchers quickly took much of the land. The nature of the environment dictated that these would be large operations and railroad towns such as Marfa and Marathon would become major shipping points. But not all cattle were adapted to these grassland mesas divided by mountainous terrain.

The Highland Hereford Association was organized in 1919 to promote the cattle business in the region and especially to tout the advantages of the particular breed of Herefords produced locally. In its literature the association took a promotional stance: "When the Big Bend County was created a program of lavish extravagance prevailed. It is one of the most remarkable sections in the Southwest—an area filled with violent contrasts. It is a storehouse of geological mystery that has fascinated and challenged man's skill to reveal and reconstruct its making by studying its marvelous formations."

Besides echoing the sentiments of Ross Maxwell, the association was prophetic in its statement that the Big Bend "charms and thrills all beholders and is destined to become the play ground of countless thousands of nature-admiring, beauty-loving, pleasure-seeking vacationists." Even as the association published its history, *The Story of the Highlands,* there were plans afoot to protect and preserve the bend of the river that would eventually involve Texas' foremost newspaperman and both the state and federal governments.

The creation of Big Bend National Park is as convoluted and complex as any activity in this country that has to do with public money and private lands. It is a story so complex that an entire book has been written about the creation of this one park. John Jameson's *Big Bend on the Rio Grande: Biography of a National Park* tells the tale, and, like the creation of many of the nation's parks, it involved years of work, numerous characters, and politics as usual.

Everett Ewing Townsend, a state representative from Brewster County, was the instigator and, until his death, pushed the process long and hard. Townsend and R. M. Wagstaff, a state representative from Abilene, guided the creation of Texas Canyons State Park, May 27, 1833. Major assistance in this endeavor came from Texas folklorist J. Frank Dobie and W. D. Smithers, who produced a set of lantern slides to promote the park. A special session of the legislature that fall provided additional acreage, and the name was changed to Big Bend State Park.

The next step for Townsend was the creation of a national park. At the time there was no national park in Texas. Lobbying the district congressman, Ewing Thomason, Townsend pushed for national park status. He took the lead in digging a well in the Chisos Basin and then building a road in order that the Civilian Conservation Corps could establish a camp there to build infrastructure. When the National Park Service reported favorably, Texas senators Tom Connally and Morris Sheppard introduced legislation, which passed and was signed by

President Franklin Roosevelt. The wrinkle in the deal was that the State of Texas was responsible for acquiring the land.

In 1937 the Texas Legislature approved $750,000 for the purchase of the private lands that would be required to complete the park. Governor James Allred promptly vetoed the measure, claiming that funds were not available. At that point Amon G. Carter, publisher of the Fort Worth *Star-Telegram* and a leading collector of western art, entered the fray. At the time the *Star-Telegram* was the premier regional paper west of Fort Worth, and it provided strong support for a park. Carter was elected chair of the newly formed Texas Big Bend Park Association. The association and Carter enlisted support from other newspapers, asked for contributions large and small from throughout the state, and successfully lobbied the legislature for an adequate appropriation to seal the deal. On September 5, 1943, Governor Coke Stevenson presented the deed of land to Amon Carter who, in turn, gave it to President Roosevelt on July 6, 1944. With the appointment of Ross Maxwell, superintendent, the park was declared open six days later.

Big Bend National Park now encompasses 1,250 square miles that includes over 100 miles of the Rio Grande. To these protected lands the state in 1988, through Texas Parks and Wildlife, added another 300,000 acres with the purchase of Big Bend Ranch. In 2008 the Texas Nature Conservancy purchased a parcel known as the Norsworthy tract that was then sold to Texas Parks and Wildlife, providing further access and protecting two significant sections of the Rio Grande. Today Big Bend State Park is the largest unit in the state park system.

My first encounter with the Big Bend was shortly after the park was officially dedicated in 1955. My uncle, originally from Chicago and a transplant to Texas because of the war, loaded his son and me in the car and took off to see what all the fuss was about. I remember arriving in the Chisos Basin and promptly having a flat tire. There were no services, and flats were common in those days. The drive out to Alpine was of greater interest than the scenery. But the point was made and the Big Bend would become a major tourist destination, as it still is.

While ranching is important and will be for the foreseeable future, the new economy for the Big Bend and Davis Mountains region is tourism. When our visitors arrive, they find themselves in one of the most unique ecological regions in North America, the Chihuahuan Desert. The largest desert in North America, the Chihuahuan covers most of the Mexican states of Chihuahua and Coahuila and as far south as San Luis Potosi. In the United States it reaches into New Mexico and through far West Texas. The numerous small mountain ranges that reach up from the surrounding plains make this desert unique. No matter from what direction one approaches, ranges such as the Davis Mountains seem to suddenly rise up without warning. These "sky islands" with their varied climate and biotic communities establish the Chihuahuan Desert as a distinct environment. And no, we do not have saguaro cactus, an indicator plant of the Sonoran Desert to the west.

The visitor traveling the roads of the Big Bend can quickly discern this pattern of basin, or plain, suddenly broken by discrete mountain ranges. However, it is through the photographic work of Paul Chaplo that resident and visitor alike can appreciate the majesty of the Chihuahuan Desert.

Every human hand laid upon this land has brought change, which has accelerated with each group or culture that has passed this way. This is a process that will likely continue. Concerned about our changing and fragile environment when we began to create a museum exhibit, as we recently did at the new Museum of the Big Bend, defining the interpretive goals, the story we wish to tell, is always the first step. These goals may be directly expressed or can become common threads that are woven throughout the exhibit and allow visitors to seek their own conclusions. The latter was the direction we chose.

The overarching goals in our *Legacy of the Big Bend* exhibit are three, and they are pertinent to the matter at hand: first, the cultural diversity and confluence of cultures and how they are shaped by the environment; second, the fragility of the environment and the limitation of its resources; and third, the representation of the Rio Grande as both a lifeline and a political barrier. This last one has become even more important in our modern world where border protection trumps the natural flow of local residents and visitors back and forth across a river that is truly a lifeline in this desert world. From the air there is no manifestation of the river as an international boundary, but the fact that it serves as a lifeline is immediately apparent.

As I work with Paul Chaplo and study his photographs, it seems to me that these same themes may be discovered in his work. As time goes on, the pressure on the resources of this fragile environment, especially water, will increase and the changes in life along the border will become permanent. Chaplo's images capture a moment in time that may not long exist.

I was introduced to the concept of the "tourist gaze" in an essay by Dr. Sylvia Rodriguez from the University of New Mexico. Simplified to my understanding, the idea of the tourist gaze implies that as a place becomes discovered and more visitors arrive, change is inevitable. Soon development occurs and the visitor experience becomes commodified and made safe by the creation of infrastructure. This never occurs with malice or even overt intent, but sometimes we can love a place to death.

In the same manner tourists often become occupants. Some then find their surroundings too primitive and, without realizing it, begin to yearn for the necessities and amenities they left behind. Certainly not everyone falls into this mold, but even those who relish the peacefulness and lifestyle they have achieved bring pressure to bear on the resources at hand.

This is not a criticism. Change and growth are inevitable, and most of us arrived from somewhere else. It is a fact of life that this land will change both through the hand of man and the inevitable environmental and climate change the region faces. We should be thankful that Paul Chaplo has managed to capture these images at a time when life on the border is changing. Land ownership is moving away from the traditional ranching culture to recreational ownership, and increasing population and tourism put pressure on limited resources. In Paul's photographs from high above we see this land from a distance and a place where we, the viewer, will probably never be. From this perspective the land appears changeless, as opposed to what might be happening on the ground.

In another respect we need to understand that these images serve both as document and art. We know they are landforms of a place we may, or may not, have seen on the ground, but few of us will ever see them from Paul's perspective. Unless we are intimately familiar with the land, the captions are often meaningless. We then move past the names of places and, with study, we apply our own skills and knowledge to interpreting what we see. We do this every day with the world around us where everything breaks down into basic elements: line, color, texture, form, and perspective. The same universal elements of art apply, and art is a matter of personal interpretation. Ron Gleason, former museum director, always held that art was best served when there were no titles. This enables viewers to create their own interpretation, their own story. We should therefore examine Paul's work not just as documentation but also as art.

As images that often initially appear as abstract, Chaplo's work allows us to exercise our own skill and experience to interpret his work as art as well as landscape, if we desire. This provides us with a different and challenging way to see. Historical geographer and architectural historian Richard Francaviglia explains why we should view Paul's images in this way: "Interpreting landscape is akin to art history, and perhaps art criticism, for landscapes can only be analyzed in terms of the social forces and aesthetic preferences that created them. We interpret landscapes because we are intrinsically interested in their form and well as their content. Thus, we call upon two rich traditions, the artistic and the scientific, that are artificially separated in our culture. Furthermore, when we set out to interpret a particular landscape, or a group of landscapes, we become artists ourselves in that we create something, either verbal or diagrammatic, by abstracting the essence of the real world." If we are to follow Francaviglia's process, we cannot merely go from image to image with only a cursory look, but rather we begin to provide our own personal interpretation, a worthy exercise. Looking at Paul's work in this manner allows one to appreciate his work as art, for Paul's eye as a photographer goes beyond mere documentation.

Almost as soon as Daguerre perfected his process, the discussion began about photography as art. While the artist could exercise his imagination, the photographer was considered to be merely a mechanic attached to his tools. As early as the 1850s

Gustave Le Gray, an early photographer, wrote, "As for myself, I express the desire that photography, instead of falling into the realm of industry and commerce, should come into that of art. That is its only true place." This argument would continue until Alfred Stieglitz opened the Photo Secessionist exhibition and published the magazine *Camera Work* in 1902. According to Thomas Livesay, photographer and former museum director, "It was exactly this time and exactly this moment that photography forever crossed the divide between a 'mechanical process' and became art. In reality, of course, it had always been so, but not recognized as such. Stieglitz did this simply by organizing a photography exhibition that would be judged and juried on its own merits rather than by the merits of painting and printmaking." At a time when the Brownie camera, introduced in 1900, made everyone a photographer, it took Stieglitz and those that followed to establish this medium as art.

Photography exists in the moment while the painter is able to return to his subject multiple times with varying conditions. As Livesay writes, "A photograph is the mirror of reality, and immune to all but what the lens captured. The painter had the ability to edit out whatever was unattractive, or whatever failed to add to the composition." Capturing that moment is not merely taking a picture, but rather the gathering of experience, talent, control of the equipment, execution, and, yes, maybe a little luck. But talent is always foremost.

Whether we view Paul Chaplo's images in this book as documentation or art, they represent the Big Bend from a place where most of us will never be. Standing on the ground in the Big Bend National Park basin surrounded by the Chisos and looking up at Emory Peak is a magnificent sight. Looking down on the basin and the same peak in one of Chaplo's photographs provides another magnificent view, and one he has chosen to share with us.

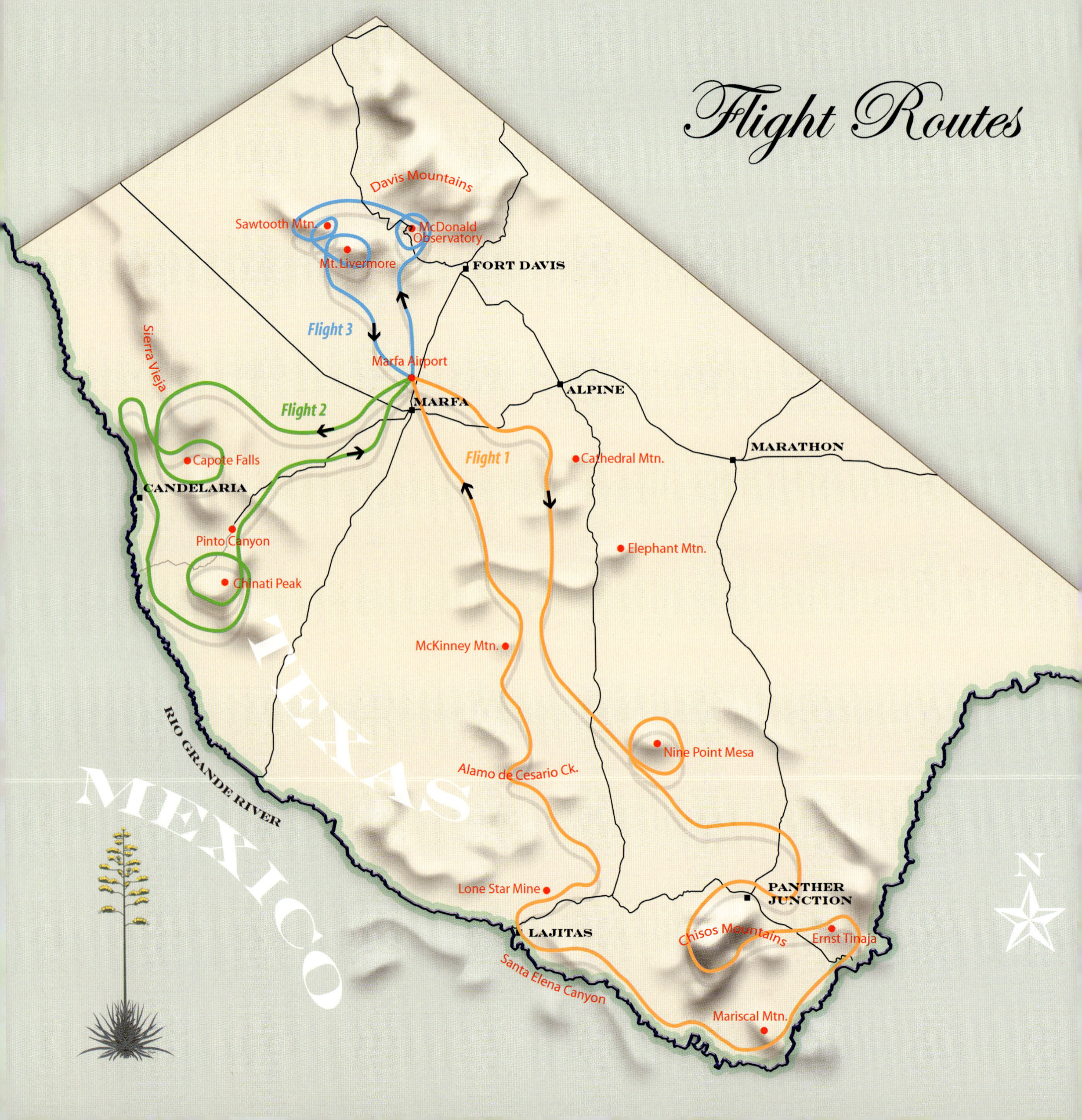
Flight Routes
Davis Mountains
Sawtooth Mtn.
McDonald Observatory
Mt. Livermore
FORT DAVIS
Flight 3
Marfa Airport
MARFA
ALPINE
Sierra Vieja
Flight 2
Capote Falls
CANDELARIA
Flight 1
Cathedral Mtn.
MARATHON
Pinto Canyon
Chinati Peak
Elephant Mtn.
McKinney Mtn.
TEXAS
RIO GRANDE RIVER
MEXICO
Nine Point Mesa
Alamo de Cesario Ck.
Lone Star Mine
LAJITAS
PANTHER JUNCTION
Chisos Mountains
Ernst Tinaja
Santa Elena Canyon
Mariscal Mtn.
N

I. THE JOURNEY BEGINS

The Flight from Marfa, Texas

Routes flown in the making of the aerial photographs in this book.
Map by author.

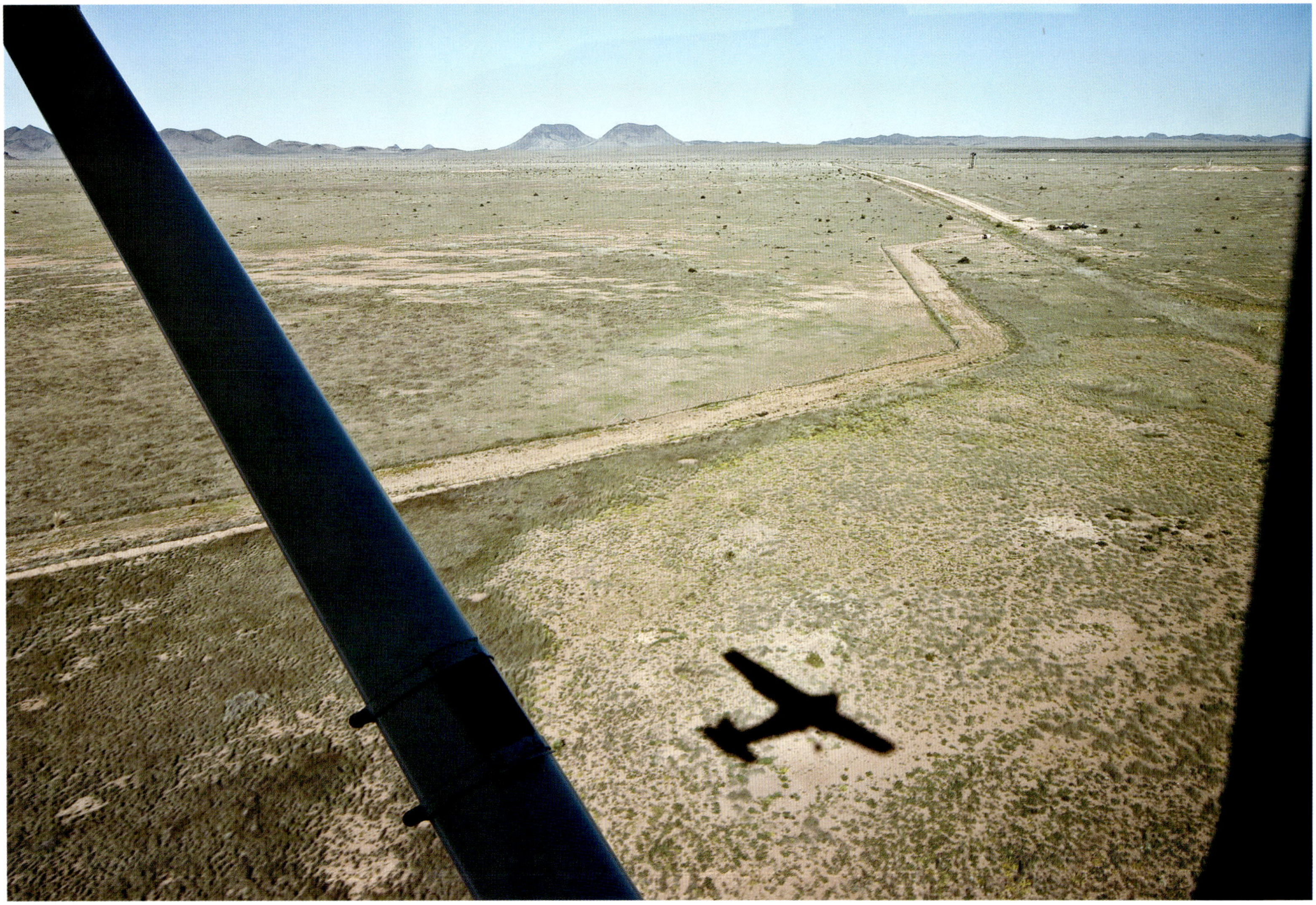

1. Marfa Flats and Haystacks

West of the Marfa Municipal Airport

The shadow of the Cessna 172 used to take all the photographs in this book glides over the desert floor below us at low altitude as we cross the Old County Road between Marfa and Bloys Camp, where the famous religious cowboy camp meetings were held. In the distance are the twin peaks known locally as the Haystacks.

2. Departing Marfa

Flight to Pinto Canyon

Flying from Marfa, we follow Pinto Canyon Road southwest across the Ryan Flat. The road appears like a ruler–straight line drawn by an engineer to make the most efficient route across this open country heading toward Pinto Canyon. In the distance are Long Draw and the Oak Hills. The open country clears my mind of Dallas worries. This is another world to me, the beginning of a journey. My old world of tall buildings and big roads is fading in the prop wash of the Cessna. Here the landscape begins to overpower the rectilinear marks of mankind. Texas Ranch Road 2810 shrinks in artifice against the canvas of the Marfa Flats.

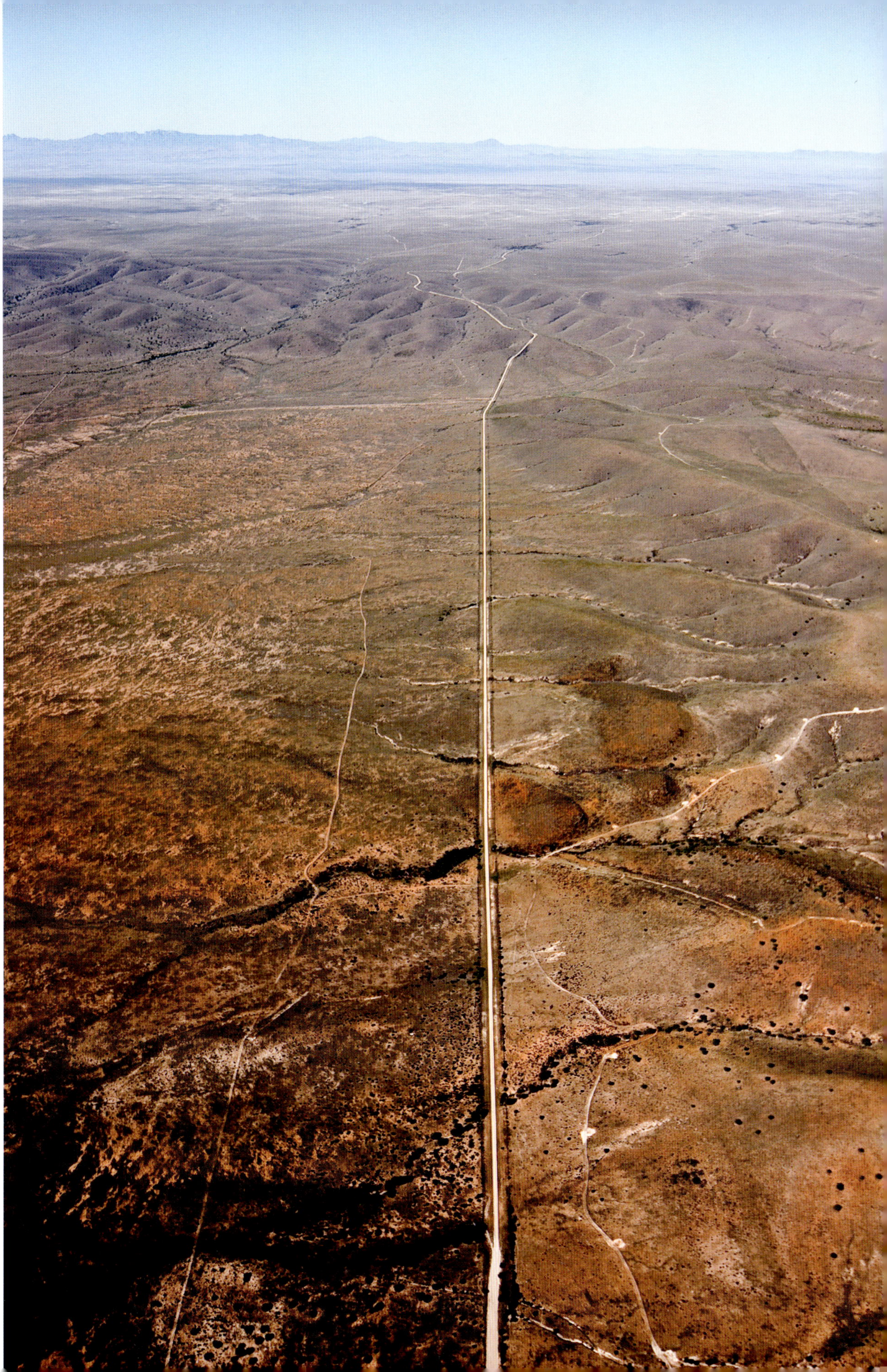

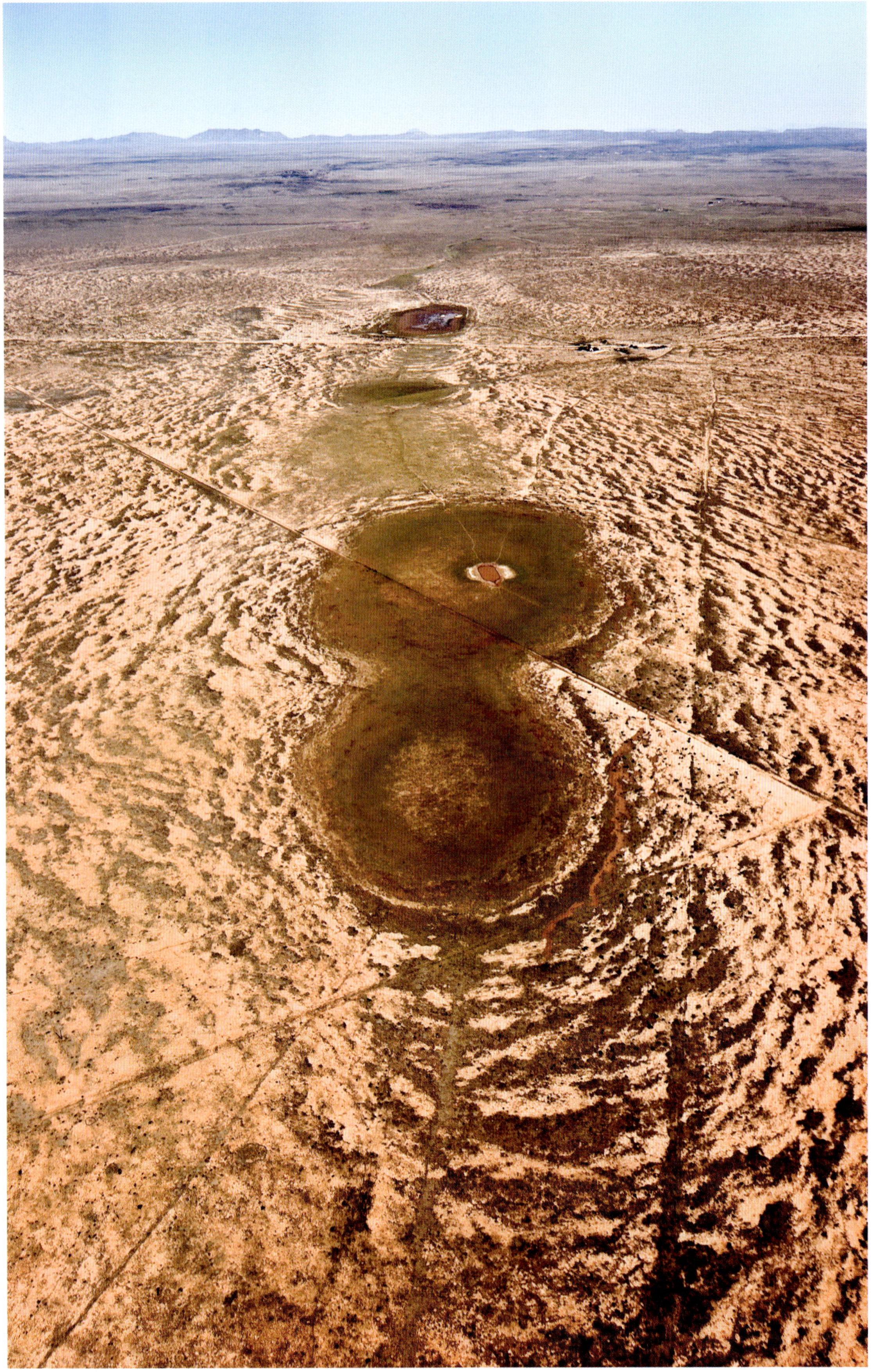

3. Marfa Yacht Club

West of Marfa

A wise person once said that to understand water, go to the desert. Near Marfa are these playas, or dried lakebeds that temporarily hold rainwater. As I see the dry earth below, I am starting to think about how I take water for granted. Some locals told me that this dry playa was the Marfa Yacht Club. I still don't know if that was an inside joke. The sharp lines of ranch roads and fences cut across these large organic pond shapes. A muddy brown water hole for cattle punctuates one of the green lake beds. The tiny lines are trails formed by the hooves of cattle wandering around what is left of the water hole. I've never been in a desert before and I learn from my pilot, much to my surprise, that Marfa gets most of its rain in the summer, especially July. It is April now and the skies are clear. In the mornings, the temperature hovers around freezing, then warms as the sun rises.

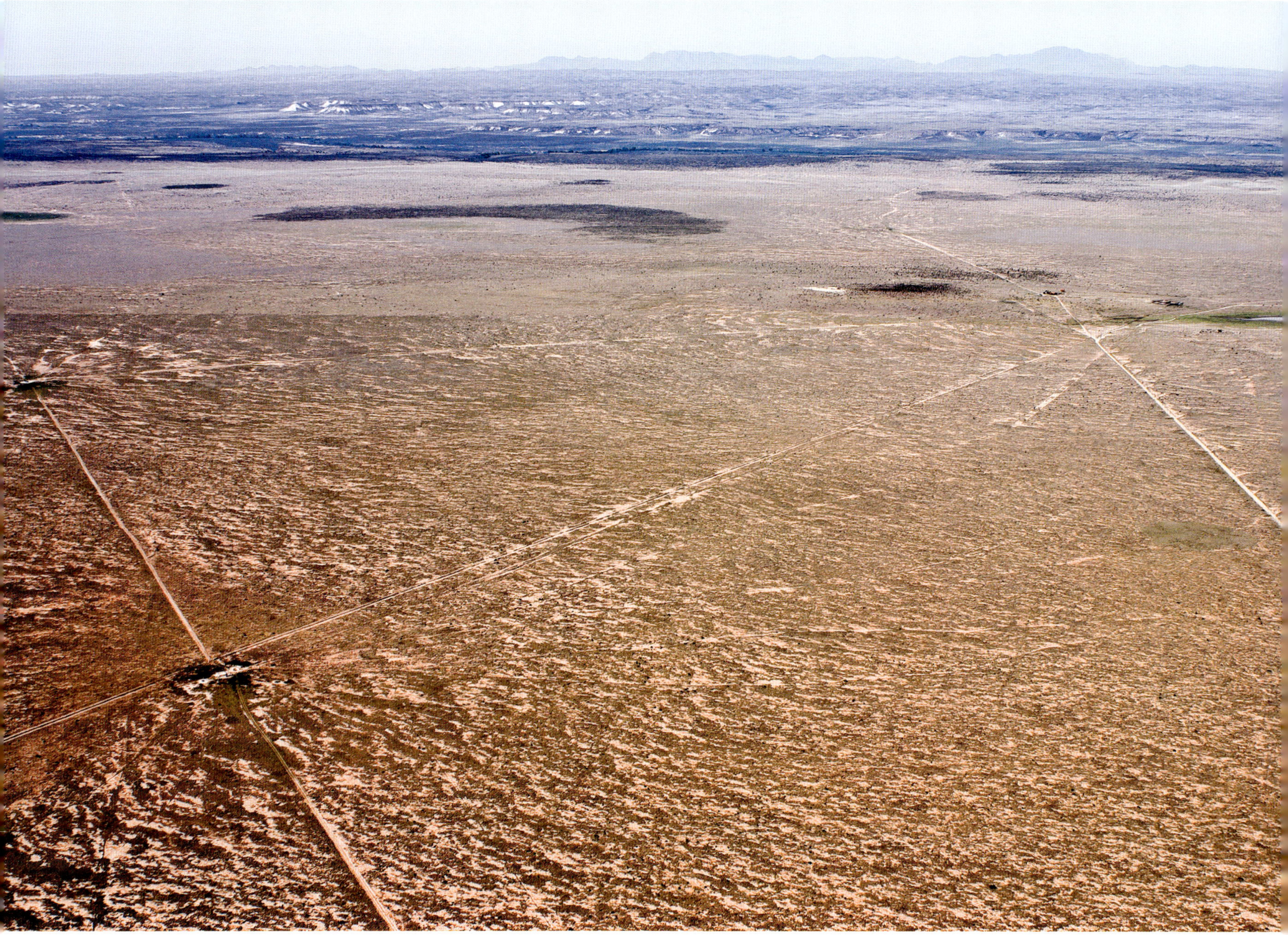

4. Marfa Flats

West of Marfa

Dirt roads geometrically divide the plane of the Marfa Flats in the ranchland surrounding Marfa. Here ranchers still raise Highland Hereford cattle, much as their ancestors did generations ago, but the country is changing—Marfa is a contemporary art outpost with national notoriety and the borderlands are the focal point of intense federal law enforcement presence.

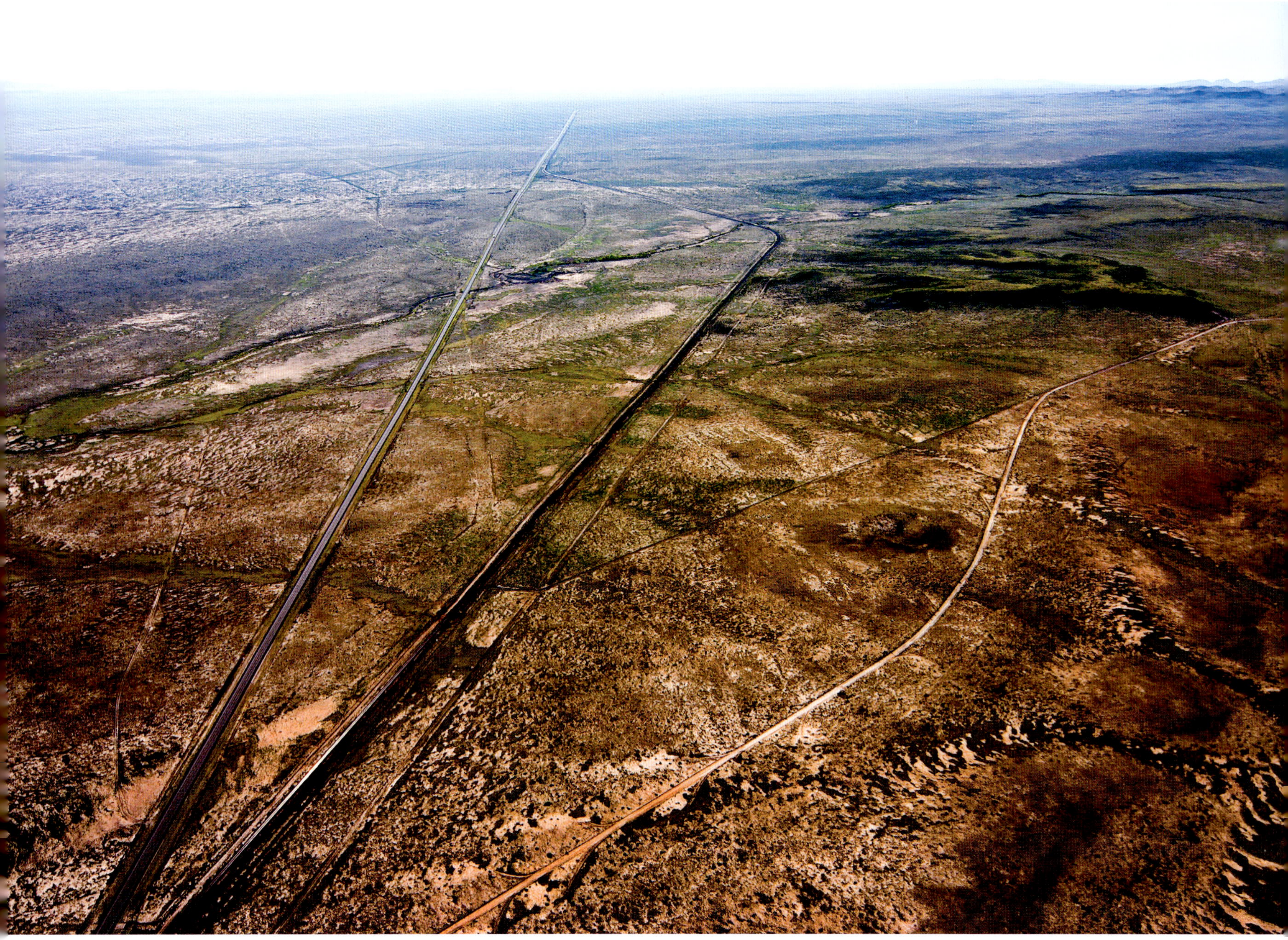

5. Diverging Road and Railroad

Marfa, Texas

Texas Highway 90 (left) makes a northwest beeline for Valentine, Texas, across the high desert. The Union Pacific Railroad runs parallel to the road, then veers away to bypass difficult terrain. Marfa was founded in the 1880s as a Southern Pacific Railroad water stop for steam locomotives. Hanna Marie Strobridge, a railroad executive's wife, named the town after a character in a novel that she had read, most likely *The Brothers Karamazov* by Fyodor Dostoyevsky.

6. Velvet Hills

Twenty-seven miles southwest of Marfa

Pinto Canyon Road cuts a graceful winding line as it follows the contours of the Oak Hills, which look like undulating velvet. Even on this approach heading southwest, the aerial view begins to reveal the varied textures of the land in Big Bend, which is foreign to my eyes. I already feel the wonder of leaving my ordinary world as my aerial "road trip" begins. I begin to snap images with my two digital cameras to get ready for what lies ahead. One camera has a normal lens, the other a very wide zoom lens.

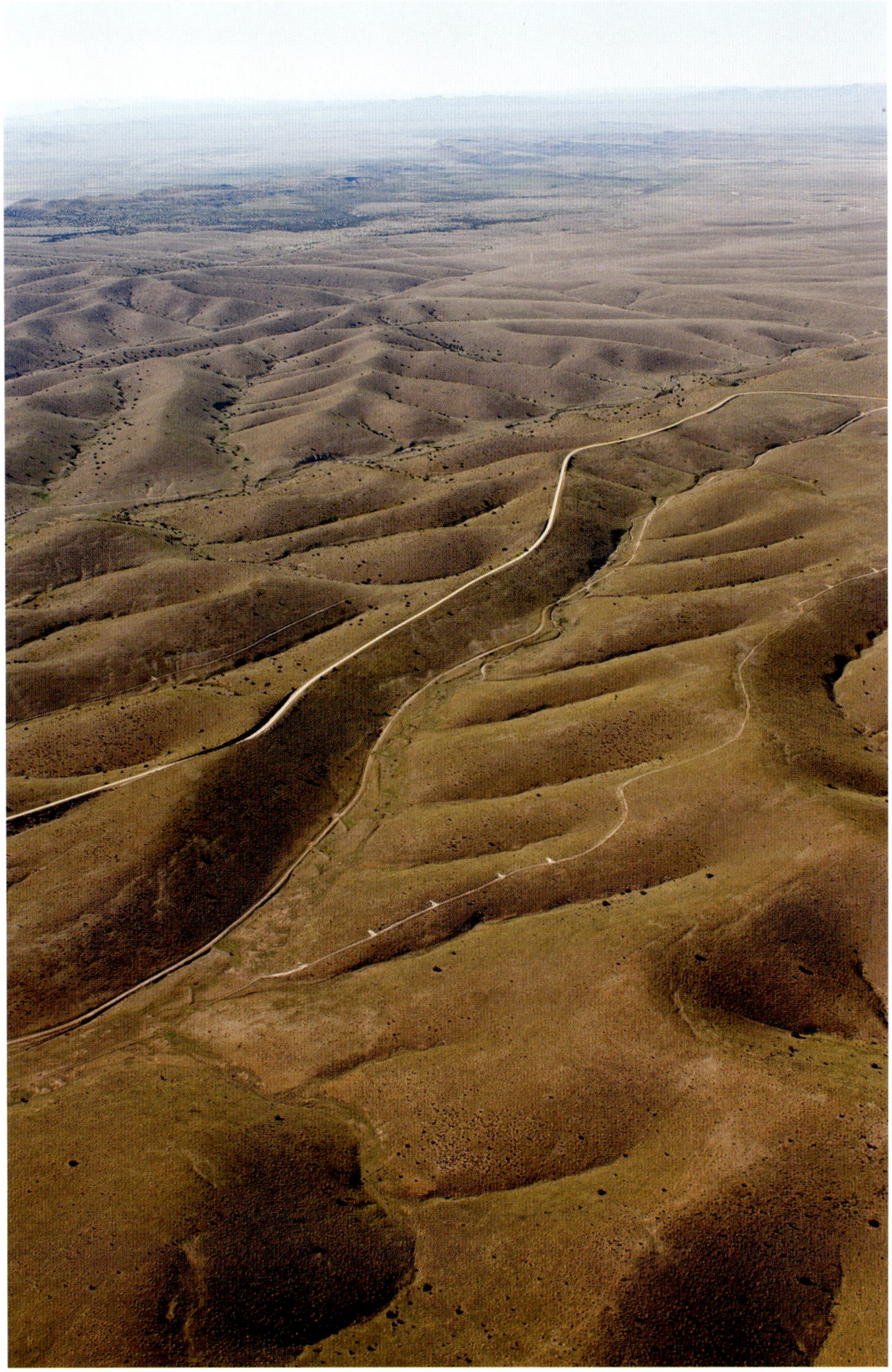

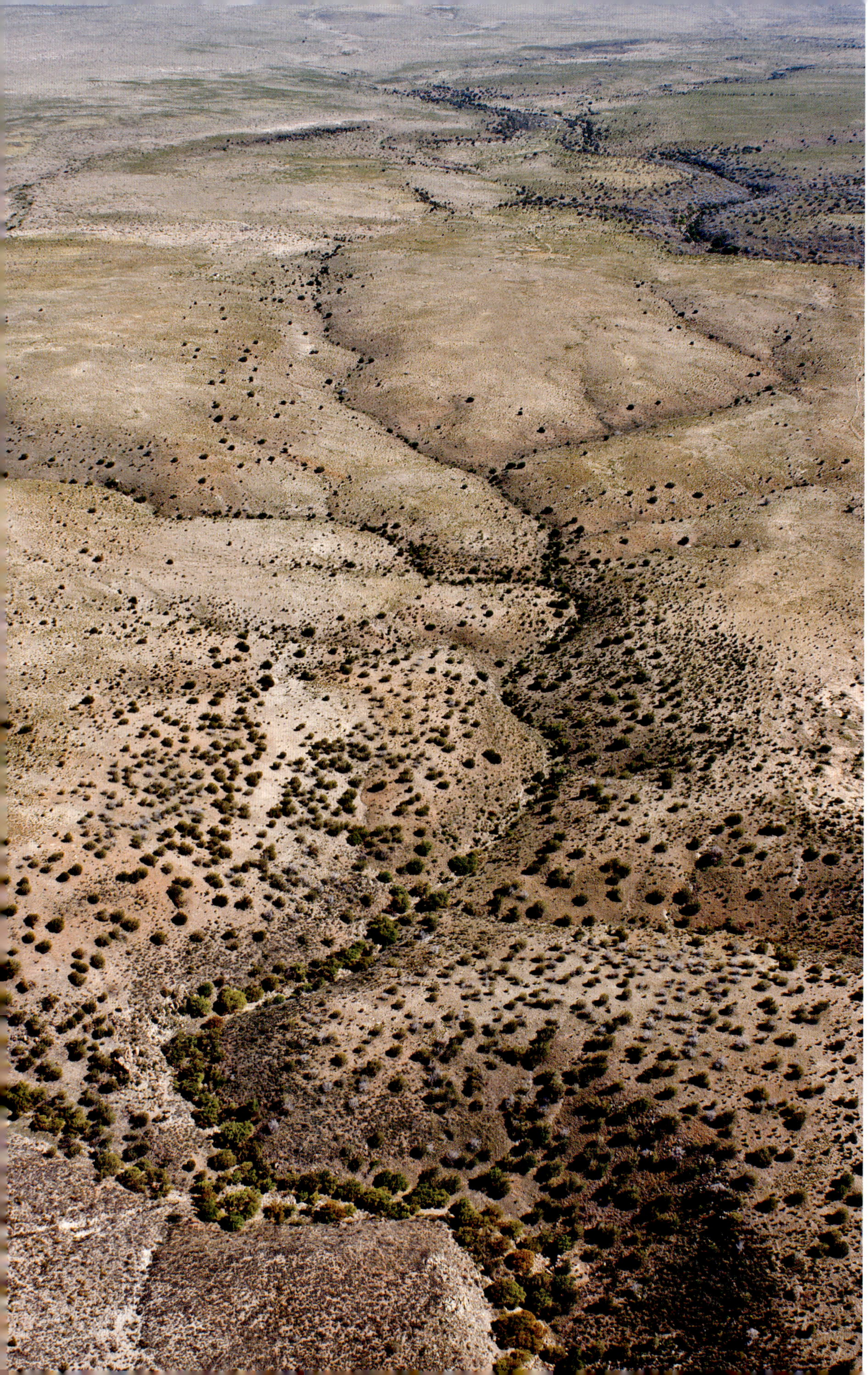

7. Juniper-Dotted Valley

Stream bed west of Marfa

Juniper and cedar congregate along a stream bed where the tributaries reach for the hills to catch the water from occasional rain showers. The green of the conifers is a striking contrast against the sandy color of the rolling hills. The distribution of the trees seems like polling results for the presence of water. Since Marfa's founding in the 1880s, ranches, which needed water for cattle, and the railroad, which required water for steam locomotives, vied for water rights.

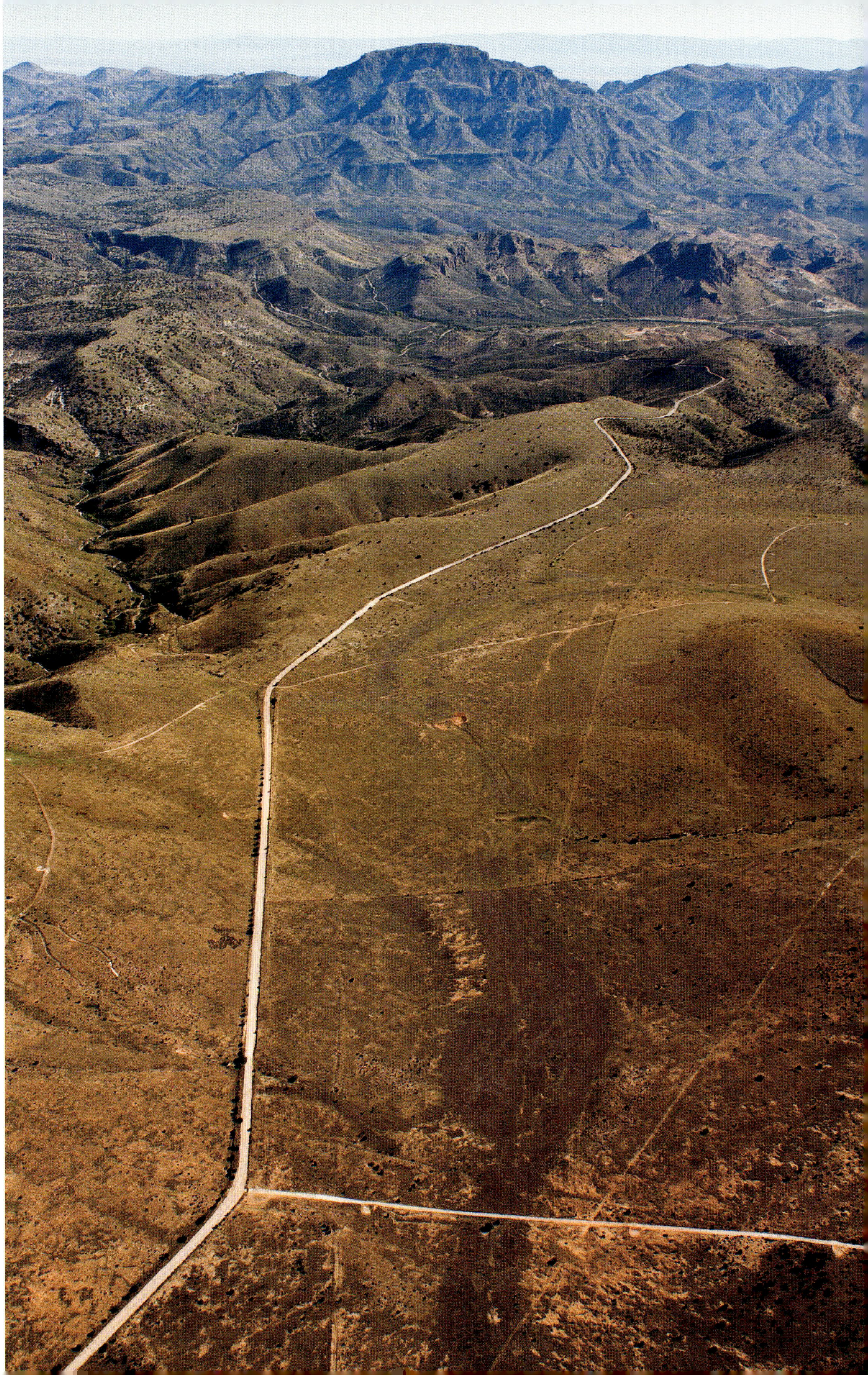

8. End of the Road

Texas Ranch Road 2810

In the foreground, the road to Petan Ranch headquarters (right) intersects Pinto Canyon Road, which continues toward the hills. Petan Ranch is still a working operation, founded by W. H. Cleveland, who ranched in the area from 1885 until his retirement at age eighty. Ahead lies an abrupt change of landscape where the paved road that sped across the flatlands becomes a winding dirt road that vanishes into the badlands. A high-clearance vehicle is recommended for the breathtaking drive beyond this point. In the distance (top, center) is Chinati Peak.

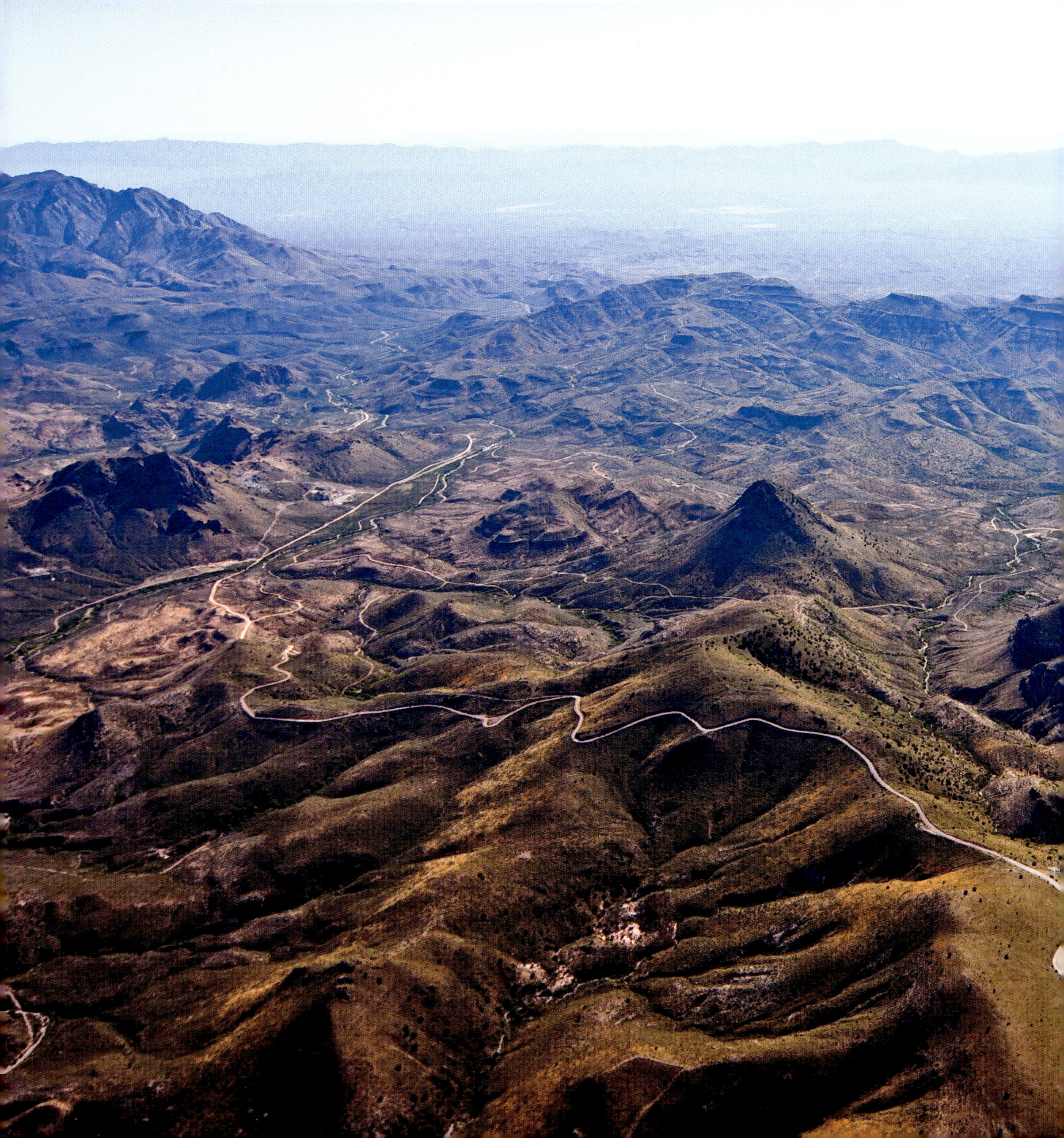

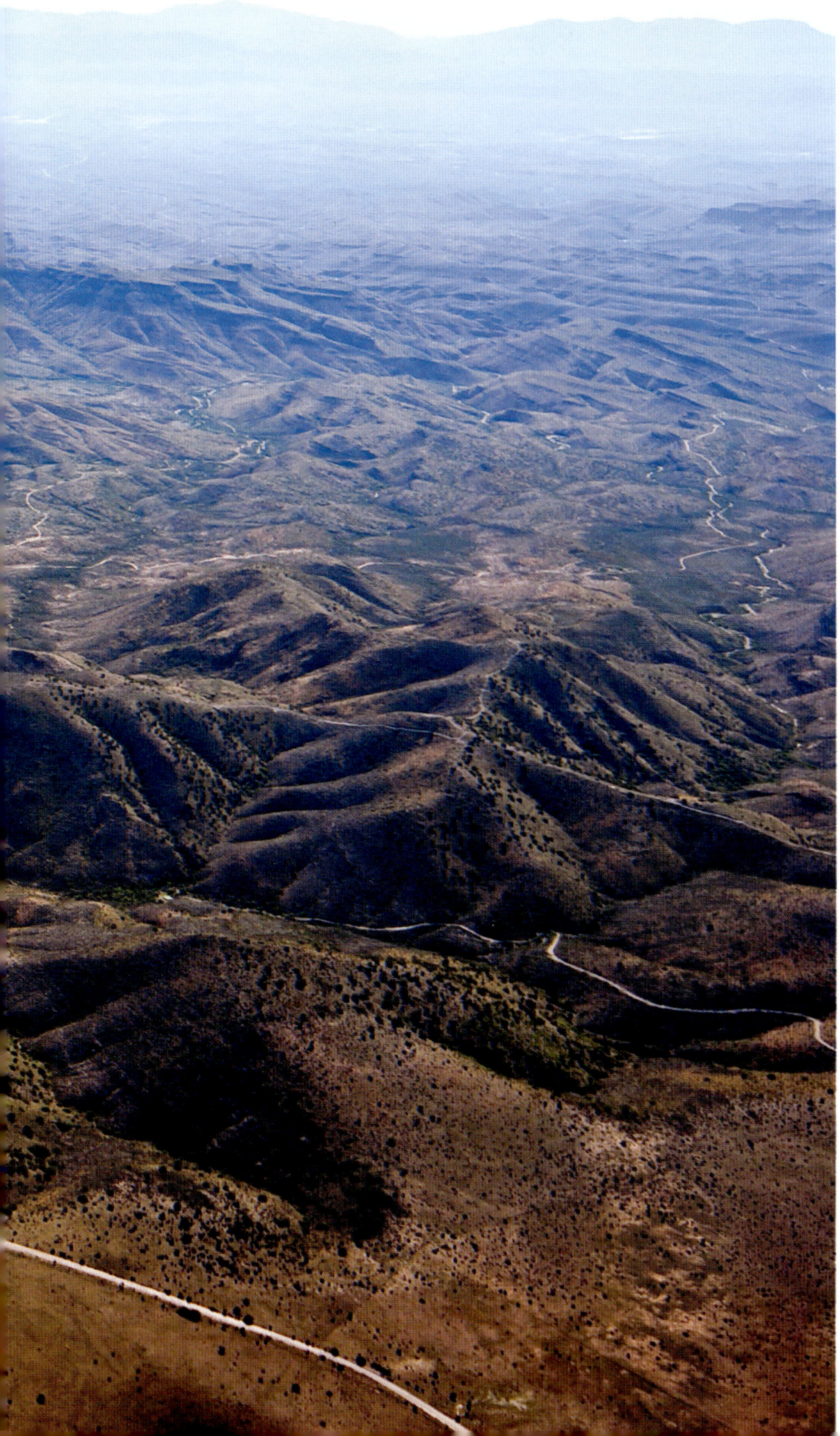

9. Into the Abyss

Pinto Canyon

Pinto Canyon Road enters the frame (lower right), then plunges off the edge of the plateau into Pinto Canyon, where it traces a serpentine path descending into a hobbit-like landscape. Farther below, to the left of where the road straightens, is the site of a defunct gold and silver mine. Just to the right of that is a ranch airstrip. Old Mexican cedars create shade along Pinto Creek in the valley. In the distance are the Rio Grande and the state of Chihuahua, Mexico. In 1914 Texas Ranger O. C. Dowe almost shot Millie Wilson, the daughter of a local rancher, while tracking smugglers in this canyon. Ranger Dowe drew a bead with his rifle, then realized that his target was a beautiful young woman. The two were married a year later, and Dowe often recounted that the first time he met his wife he almost shot her. Today Mexican smugglers, or mules, bring drugs into the United States through this rugged area. Locals advised that when I drove there to "be out by dark."

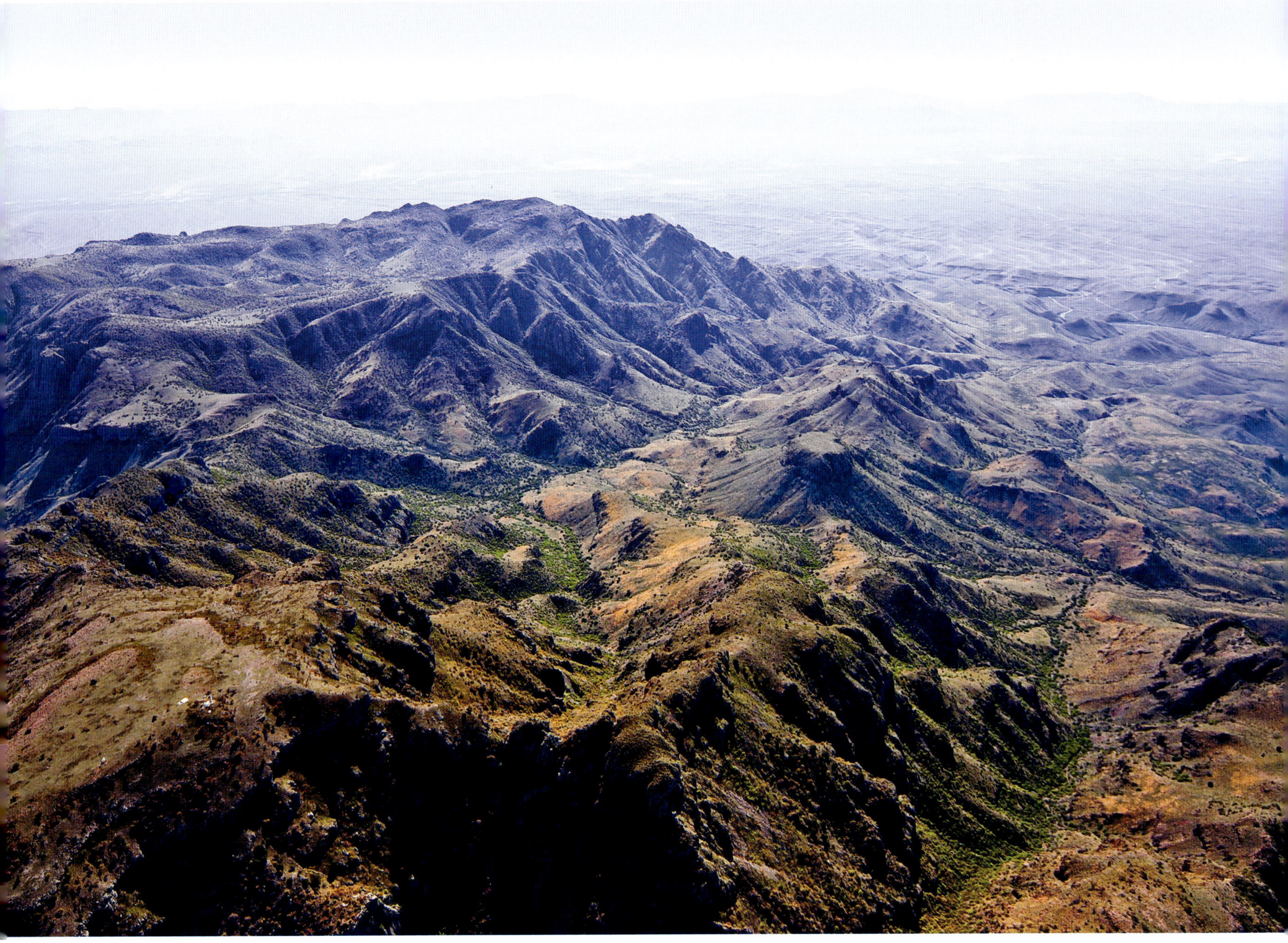

10. Chinati Mountains

East of Texas Ranch Road 2810

This scenic area is in the Chinati Mountains State Natural Area, which is Texas state land that is closed to the public. The buildings (lower left) are a law enforcement radio facility to allow communications over the mountains. A helicopter pad nearby allows the site to be serviced. The Chinati name is derived from the Apache word for mountain pass, *ch'íná'itíh.* A hint of Pinto Canyon Road is visible (far right) as it descends into the Rio Grande valley.

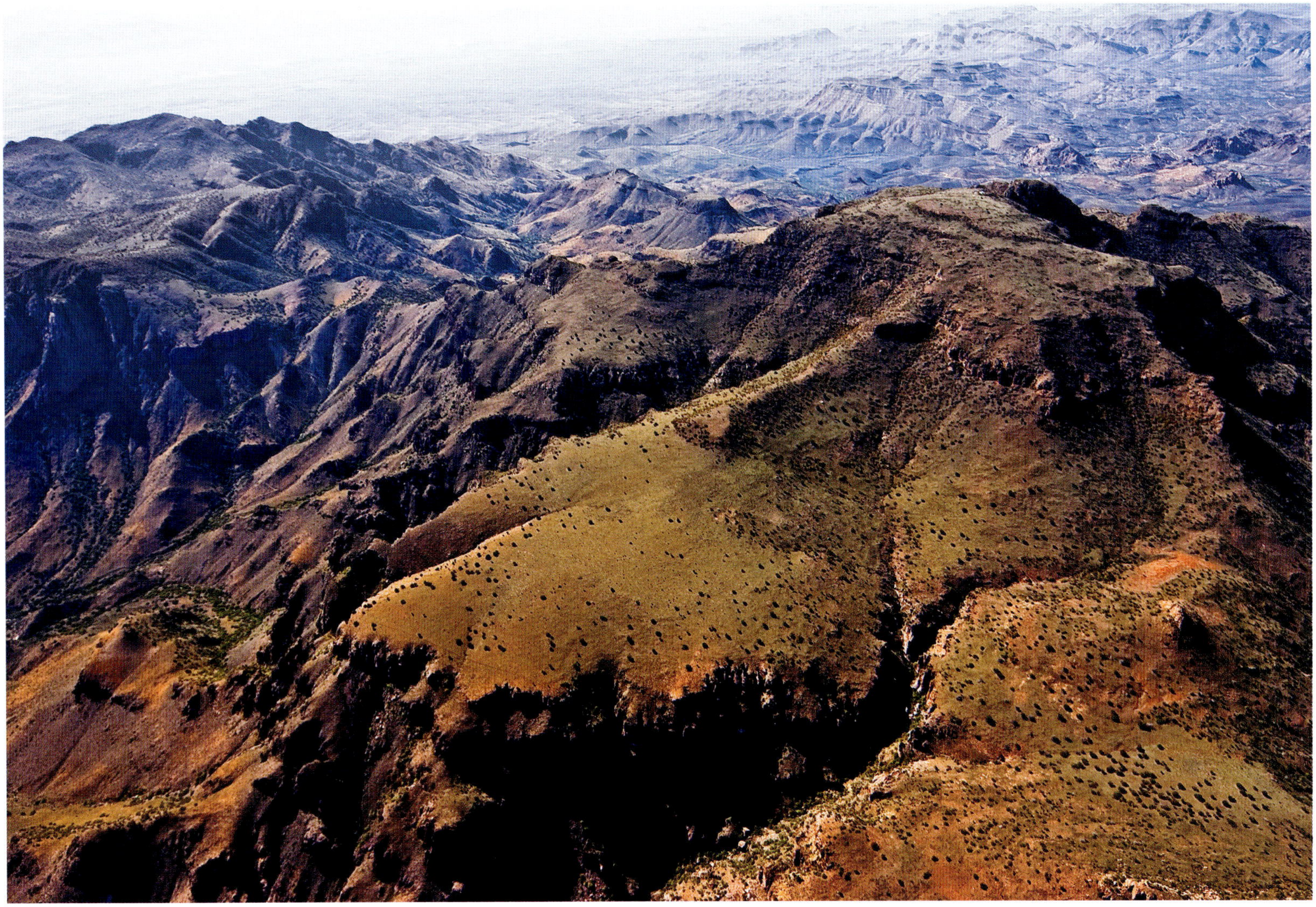

11. Chinati Peak

Twenty-five miles southwest of Marfa

Chinati Peak is the 7,723-foot snaggle-toothed remnant of the largest volcano in Trans-Pecos Texas, which was formed about thirty-two million years ago. The rock is mostly ancient lava that flowed and contorted in strange claw-like shapes visible from the air. I chose this image for the fantastic shape of the peak. The ragged edge seems like a child's seashore castle cast by dripping sand; yet here is a sleeping supervolcano that would impact the world if it erupted today.

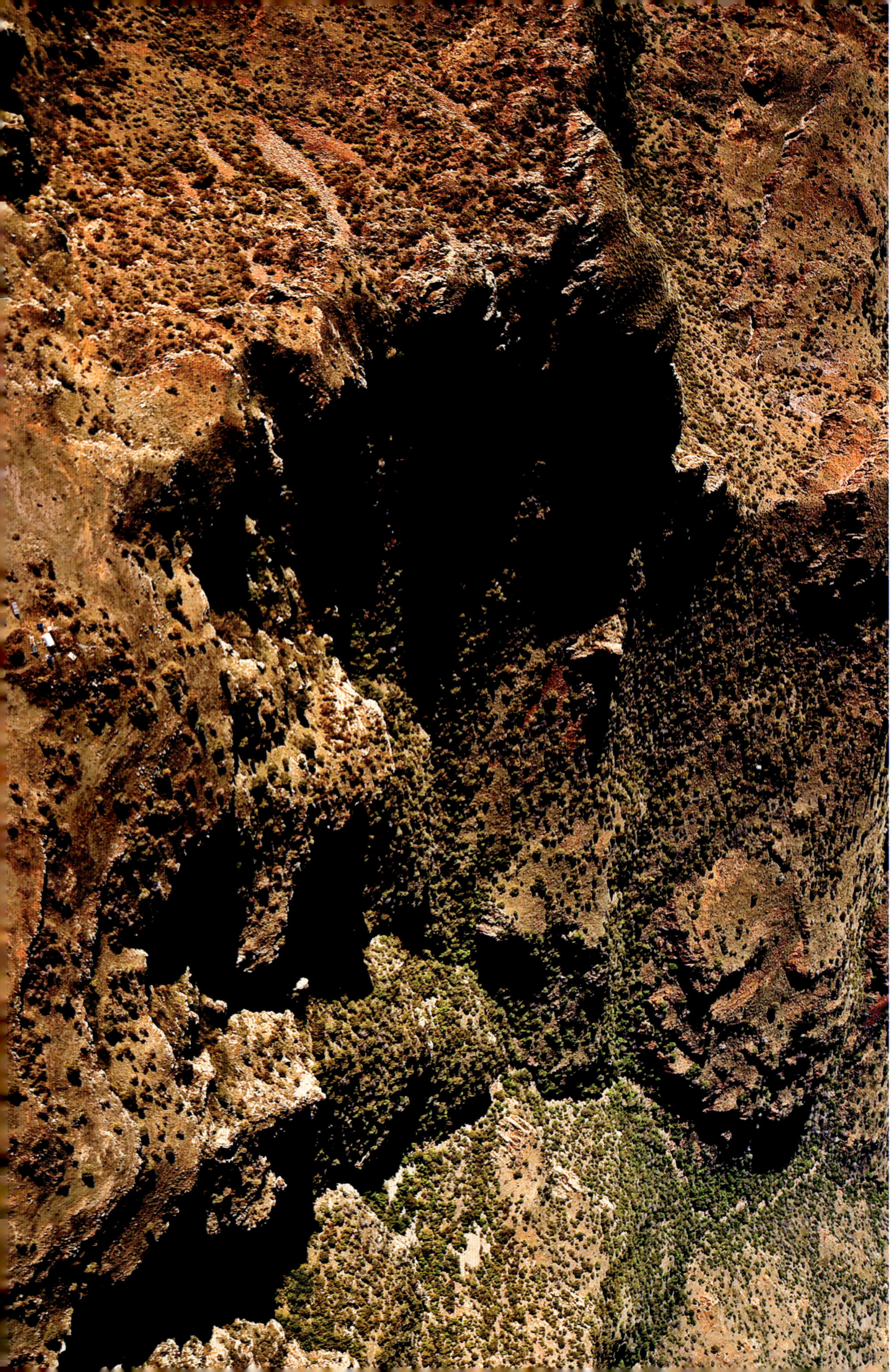

12. Precipice

Detail of Chinati Peak

The miniscule white dots are buildings that sit on the precarious edge of the massive Chinati Peak. The scale of these mountains dwarfs mankind's tiny marks as we fly over this remote restricted area where hikers are not allowed. Here is where lava flowed from a massive volcano until the chamber under the mountain was emptied, then the volcano collapsed into the hole under its own weight, creating a caldera. It is the grave of a volcano that fell into the earth. The Chinati Caldera is on the list of the world's largest supervolcanoes. I included this closeup view to show better the detail of the monstrous shape of the mountain that seems ready to swallow the tiny buildings near its gnarly crag.

13. A Thousand Mount Saint Helens

Chinati Mountains

Representing the calm after an ancient geologic storm, the Chinati Mountains are a small souvenir of the most violent volcano that ever erupted in Texas. The cataclysmic eruption sent out an estimated 262 cubic miles of superheated ash that exploded miles into the air, then raced across the surrounding land at about one hundred miles per hour, for forty miles in all directions, burning and burying everything in its path. The Chinati volcano, about thirty-two million years ago, exceeded any eruption in modern times, including Krakatau, Indonesia, and Vesuvius, Italy.

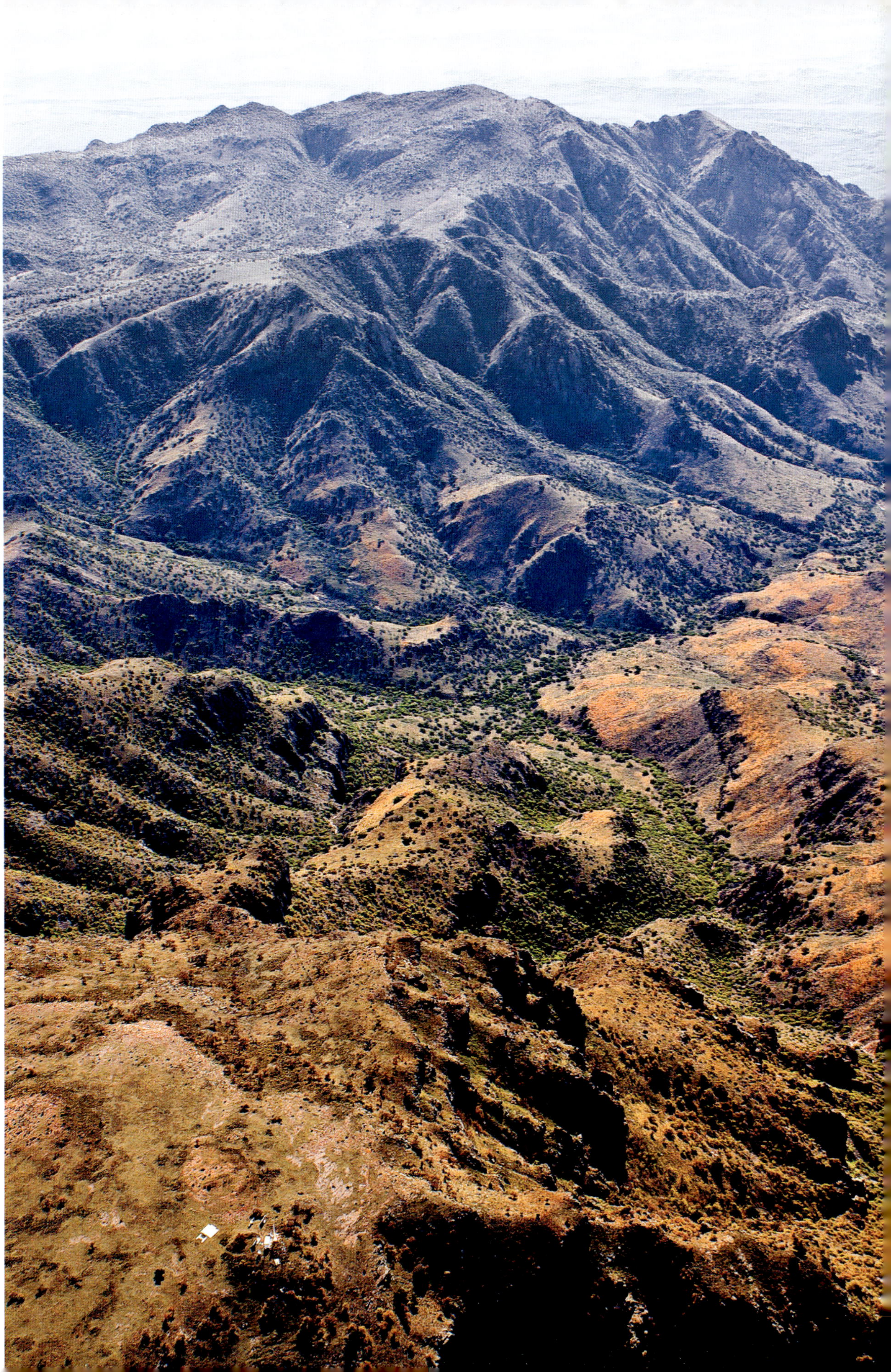

II. ALONG TEXAS HIGHWAY 118

Heading South from Alpine, Texas

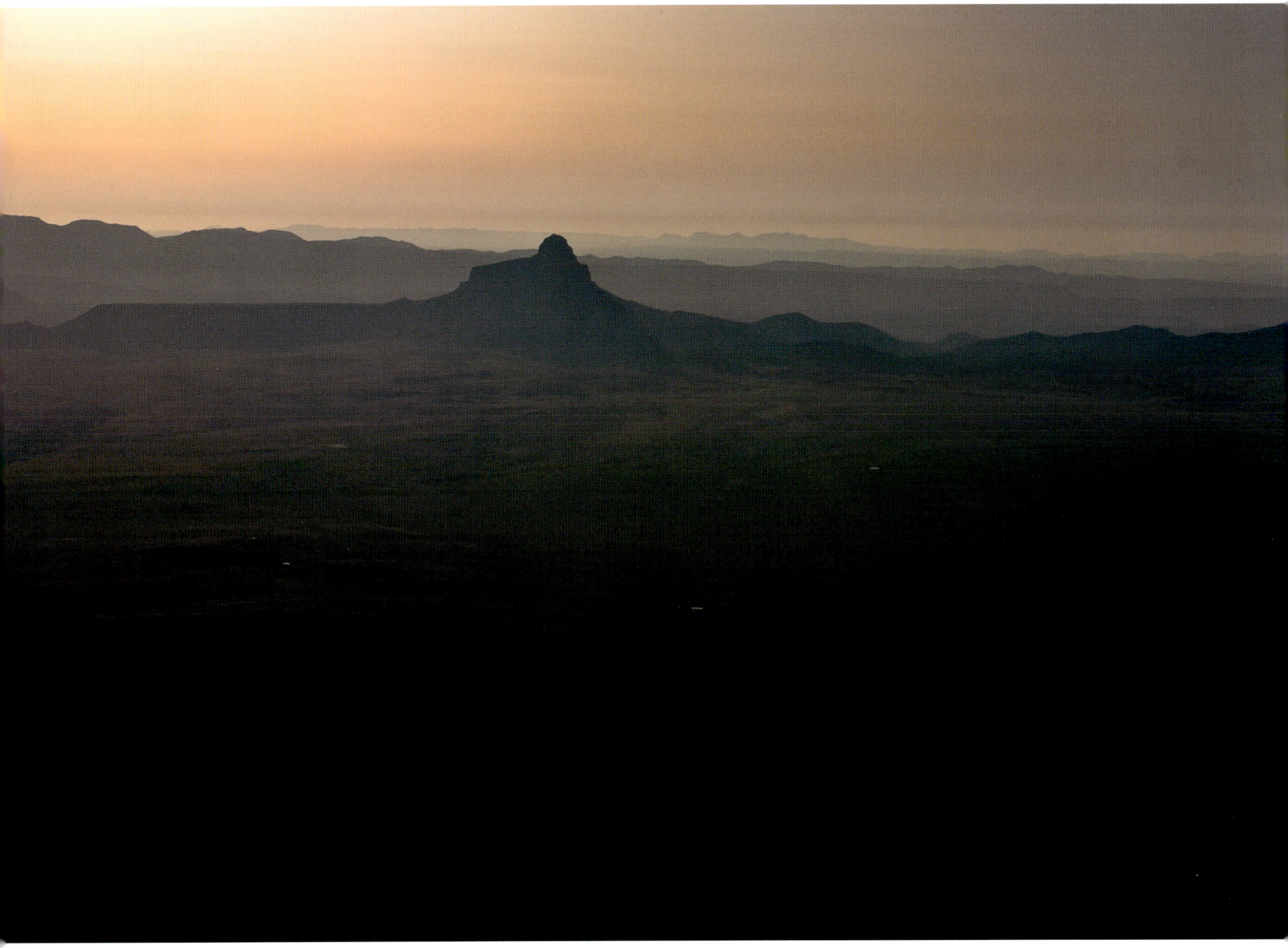

14. Cathedral Mountain Silhouette

East-southeast of Marfa

Cathedral Mountain rises like a Gothic ruin from the One-O-One Flats. Haley's Peak, the 6,800-foot summit, is named after Lawrence Haley, who operated a 15,000-head sheep ranch on 37,000 acres of land there in the mid-1880s. Early explorers described the Trans-Pecos plains as land covered with luxuriant grasses. High stocking rates of livestock, wildfires, and erosion led to rangeland degradation. Currently Cathedral Mountain Lodge operates a restaurant in this area that offers scenic views of the legendary ranch.

15. Mountain Ridges at Sunrise

North of Cathedral Mountain, looking southeast

On my first flight, we took off from Marfa in the early-morning darkness, headed east toward Alpine, then south along Texas Highway 118. As the sun rose near Cathedral and Goat mountains, I could see the layered ridges of the Del Norte Mountains to the east, looking over Calamity Creek, which takes its name from disastrous flash floods. The peaceful, dreamy mood of the dawn light of this scene appealed to me.

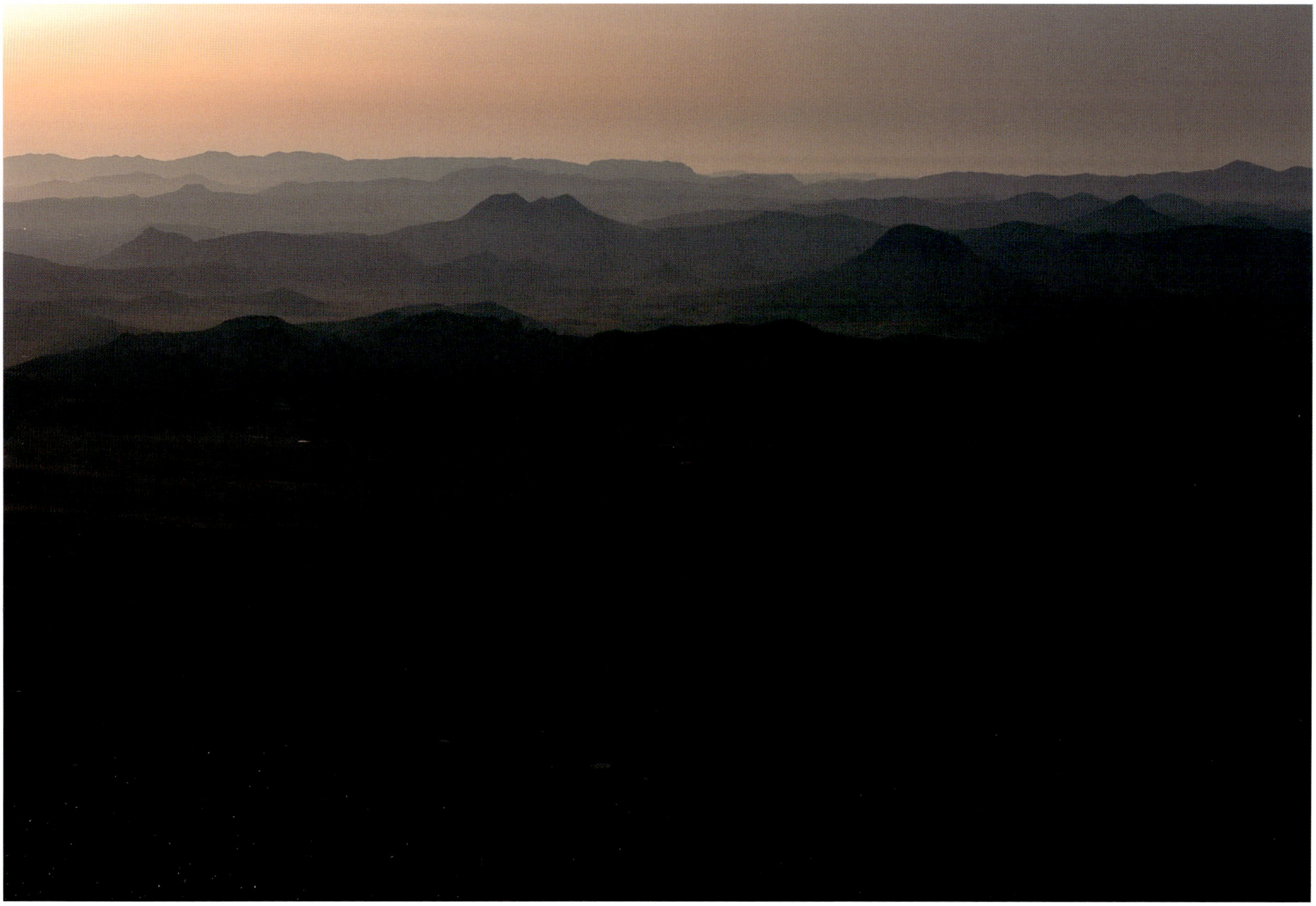

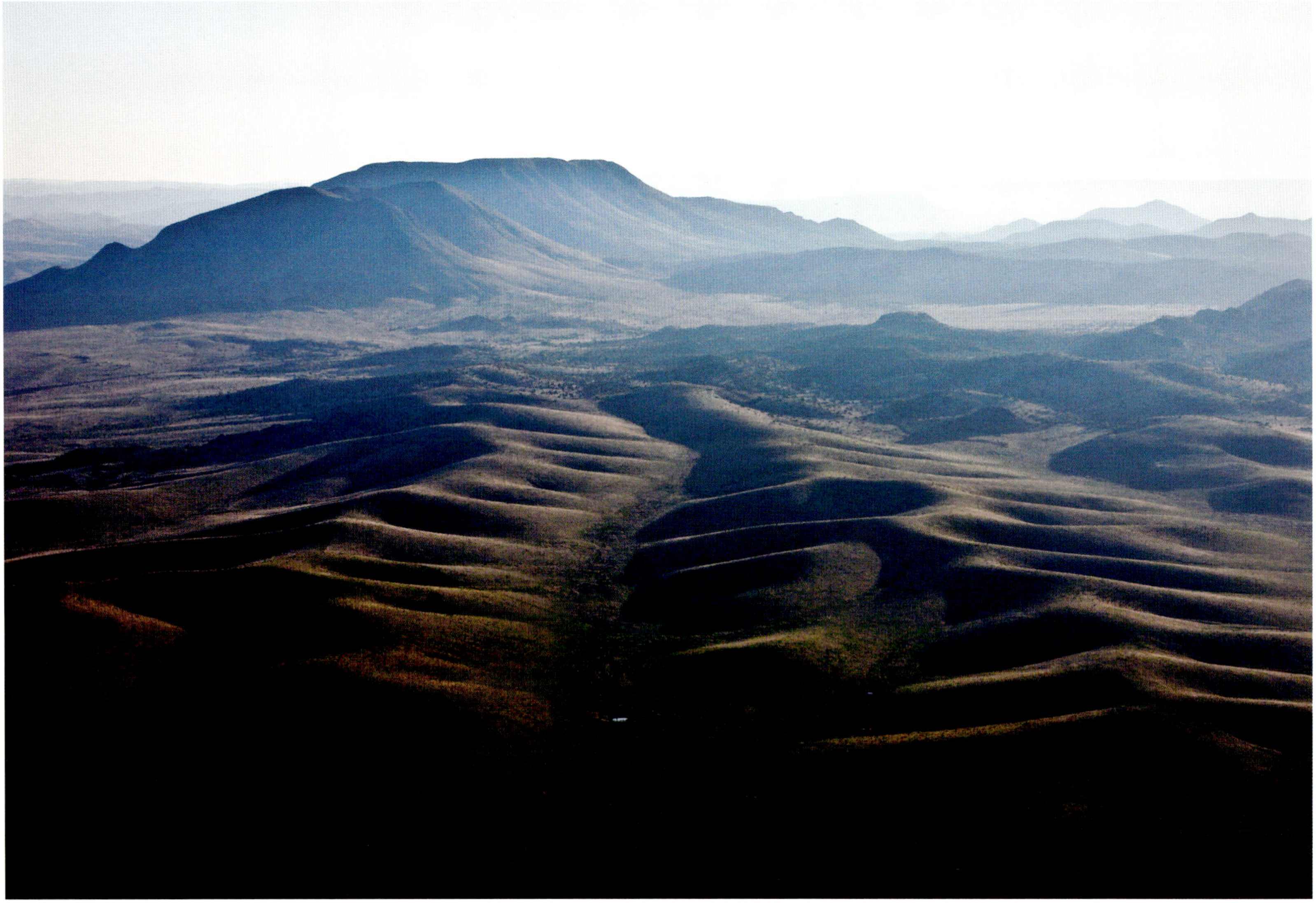

16. Goat Mountain

Southwest of Alpine

Viewed from the southeast in the morning light, Goat Mountain, a layer cake of volcanic ash and lava, rises to an elevation of 6,750 feet, about 1,750 feet above the surrounding lowlands. On the other side of the mountain is an old smugglers' route coming up from Mexico, still in use today. The low-angle light accents the soft texture of the rippled terrain as seen from the air.

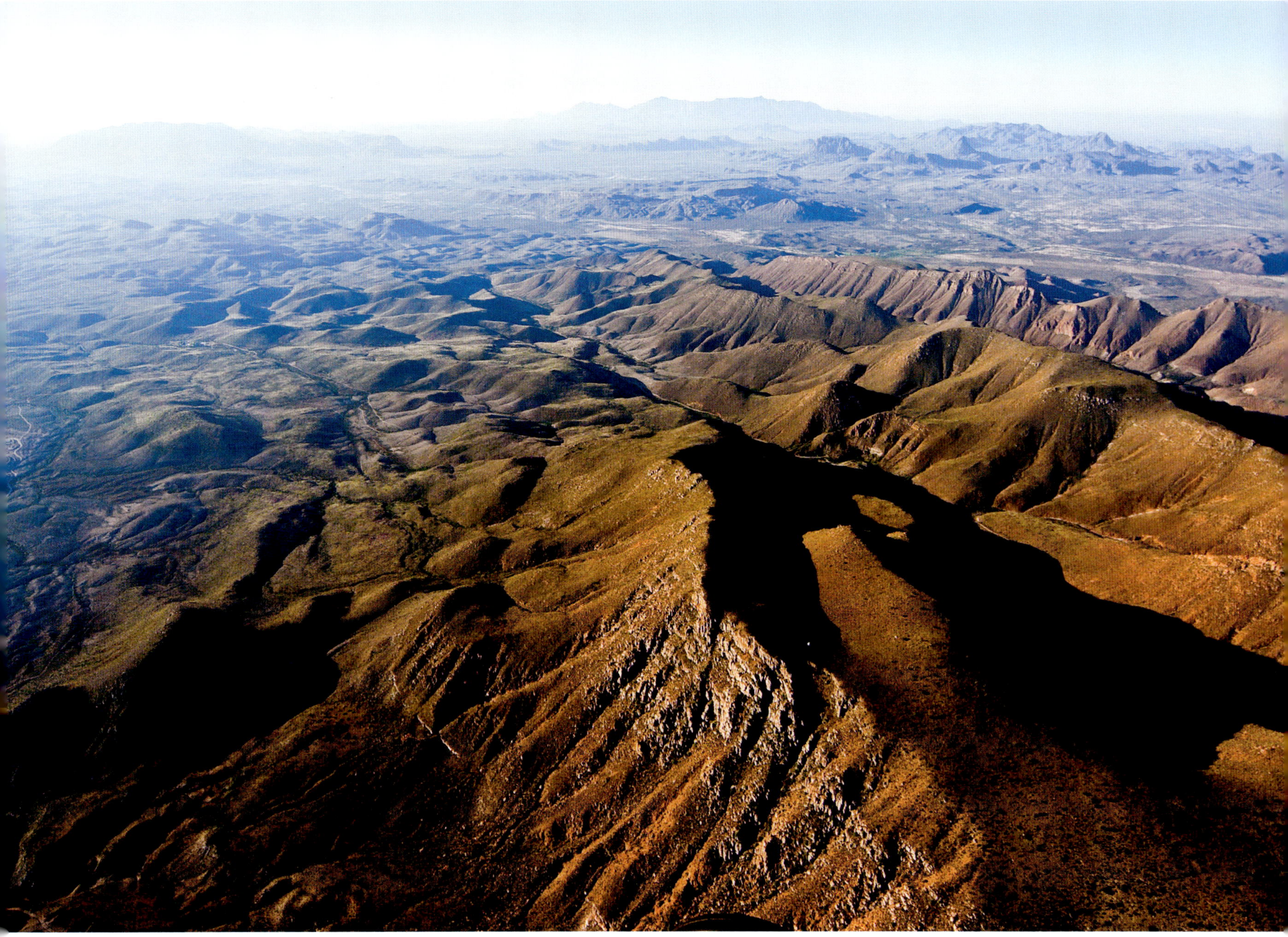

17. Canyons on Nine Point Mesa

Sixteen miles west of Persimmon Gap

The high-contrast side lighting reveals the ruggedly beautiful terrain of this remote part of Nine Point Mesa. From left to right are Pipeline Canyon and Cartwright Canyon. In the distance, looking south across Nine Point Draw, are the Chisos Mountains of Big Bend National Park. Here the ridges lead the eyes into the distance, expressing the vastness of this uniquely Texas land.

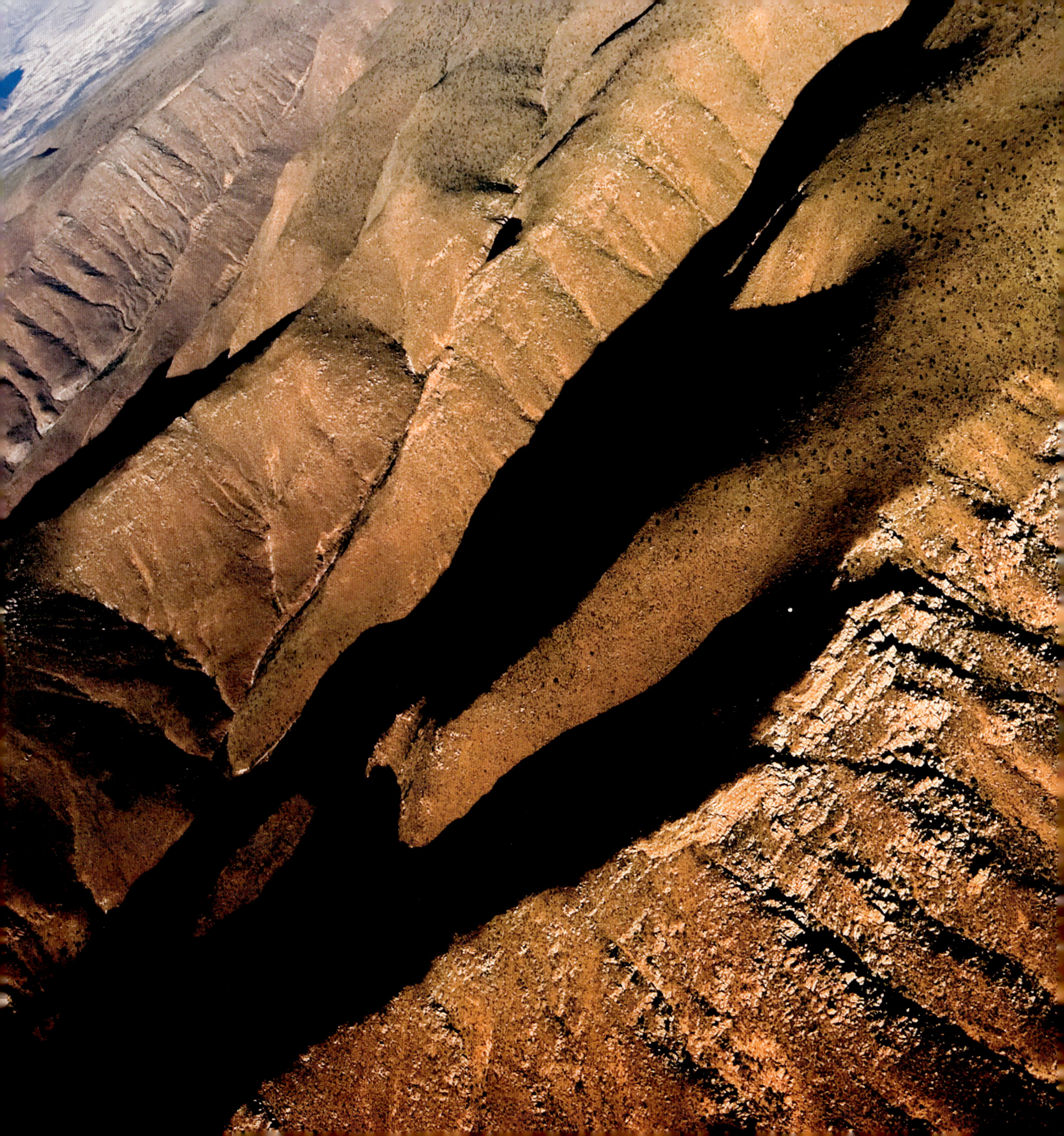

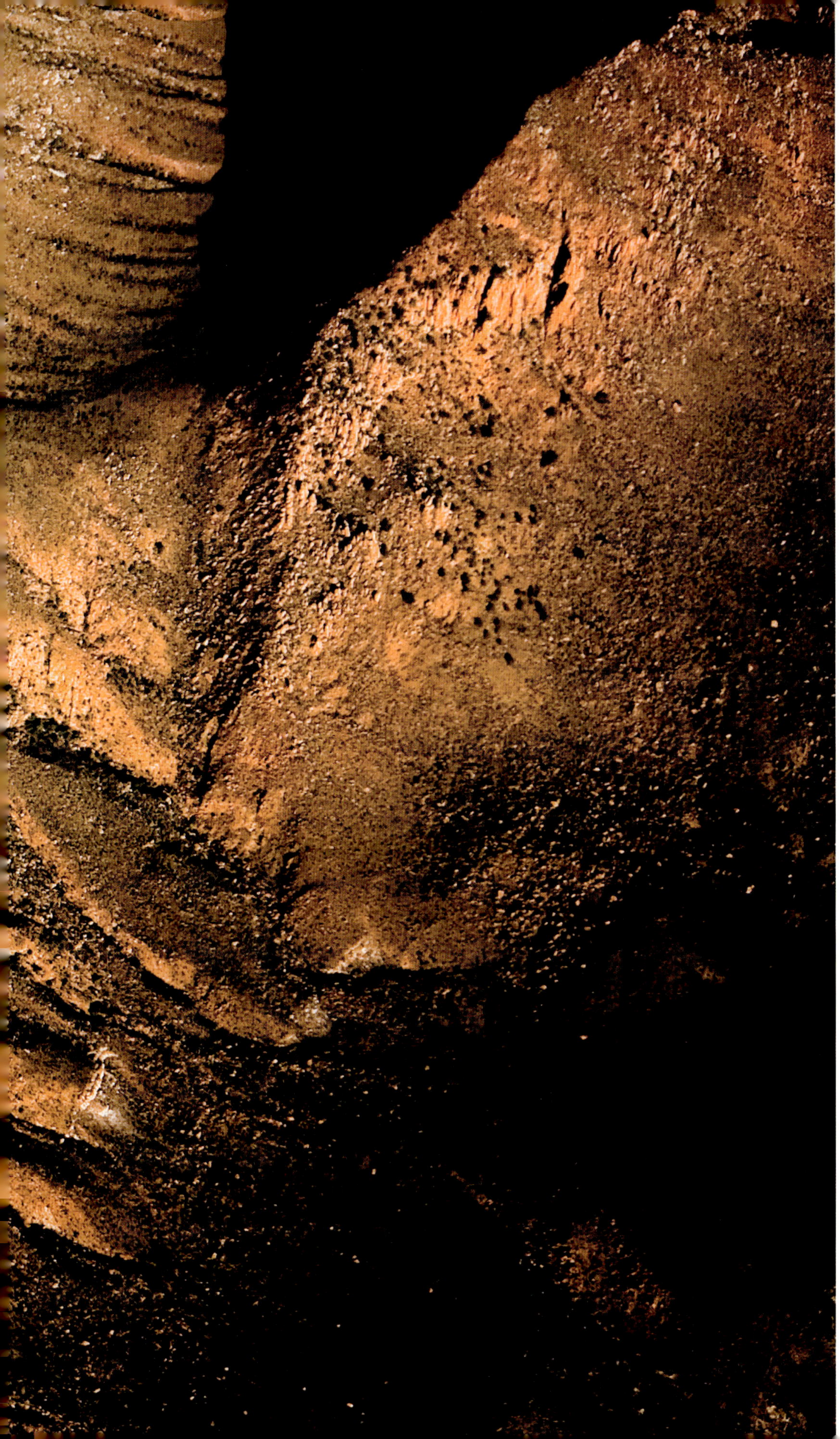

18. Looking into Pipeline Canyon

Nine Point Mesa

Away from the tourists driving on Texas Highway 118 are these dramatic escarpments and valleys on the southeast corner of Nine Point Mesa. I captured this dramatic view as we dipped the left wing of our Cessna to look almost straight down into the canyon below. Details such as this photograph are part of the visual narrative of this book, much like a close-up of a person's face in a movie, to reveal an expression.

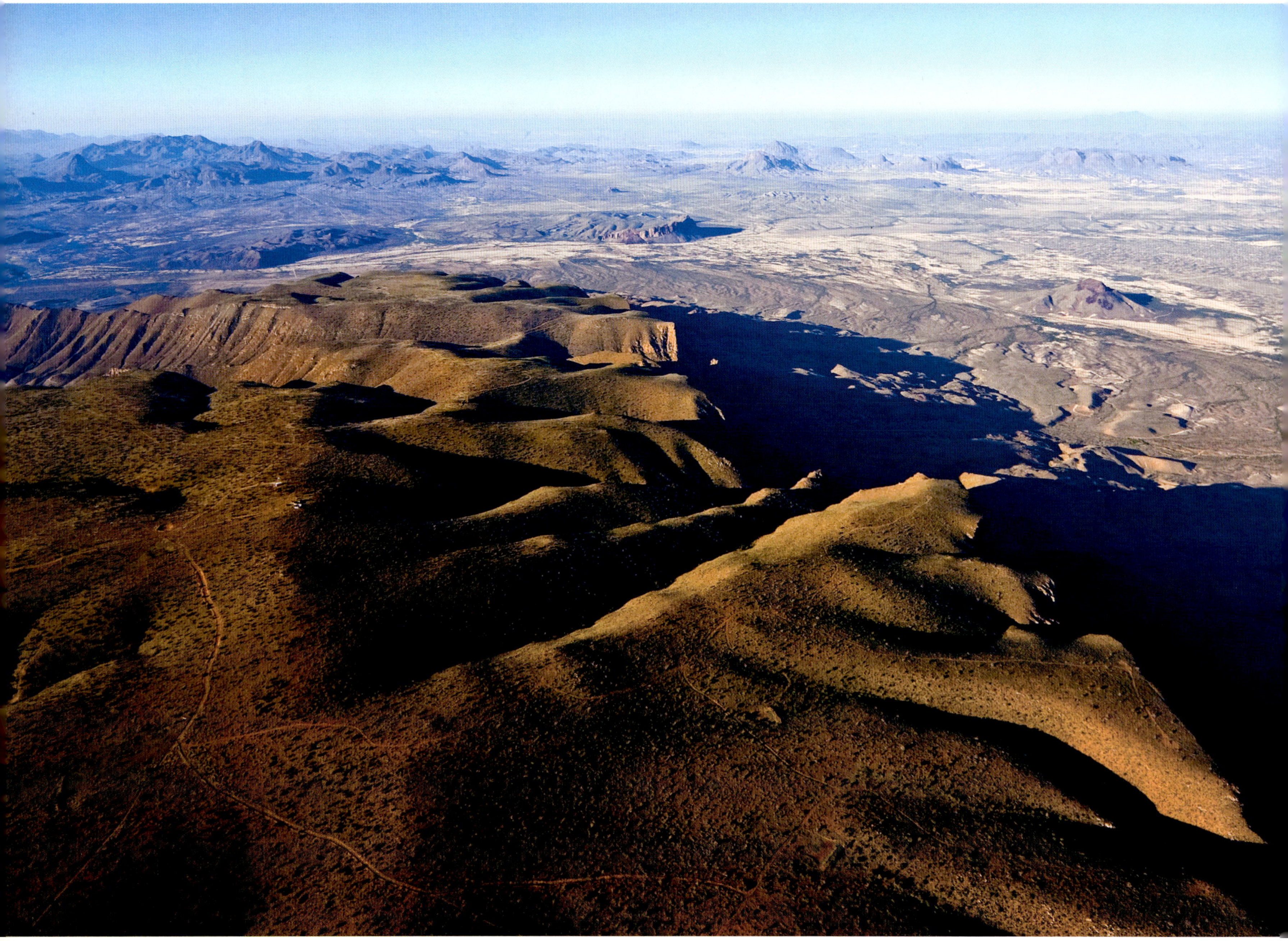

19. A New Home for Bighorn Sheep

Nine Point Mesa Ranch

In a joint venture of the Texas Parks and Wildlife Department, Sul Ross State University, the Texas Bighorn Society, and Nine Point Mesa Ranch, this mesa is being stocked with desert bighorn sheep. The animals are being captured on Elephant Mountain, thirty miles away, and then transported by helicopter. The bighorn reintroduction effort in the Big Bend region has been a particular success. I felt wonder seeing the titillating shape of the mesa top from the air, especially the ribbonlike edge.

20. Corazones Mountains

Terlingua Ranch

East Corazones Peak (center) rises 5,045 feet above East Corazones Draw, looking southwest toward the Christmas Mountains. To the left of the mountains near the shadow of Whitehouse Mountain is Terlingua Ranch Lodge and the gravel runway of the Terlingua Ranch Airport. In the farthest distance (top), twenty-seven miles away, is the downstream mouth of Santa Elena Canyon where it emerges from the cliff face of Mesa de Anguila and the Sierra Ponce.

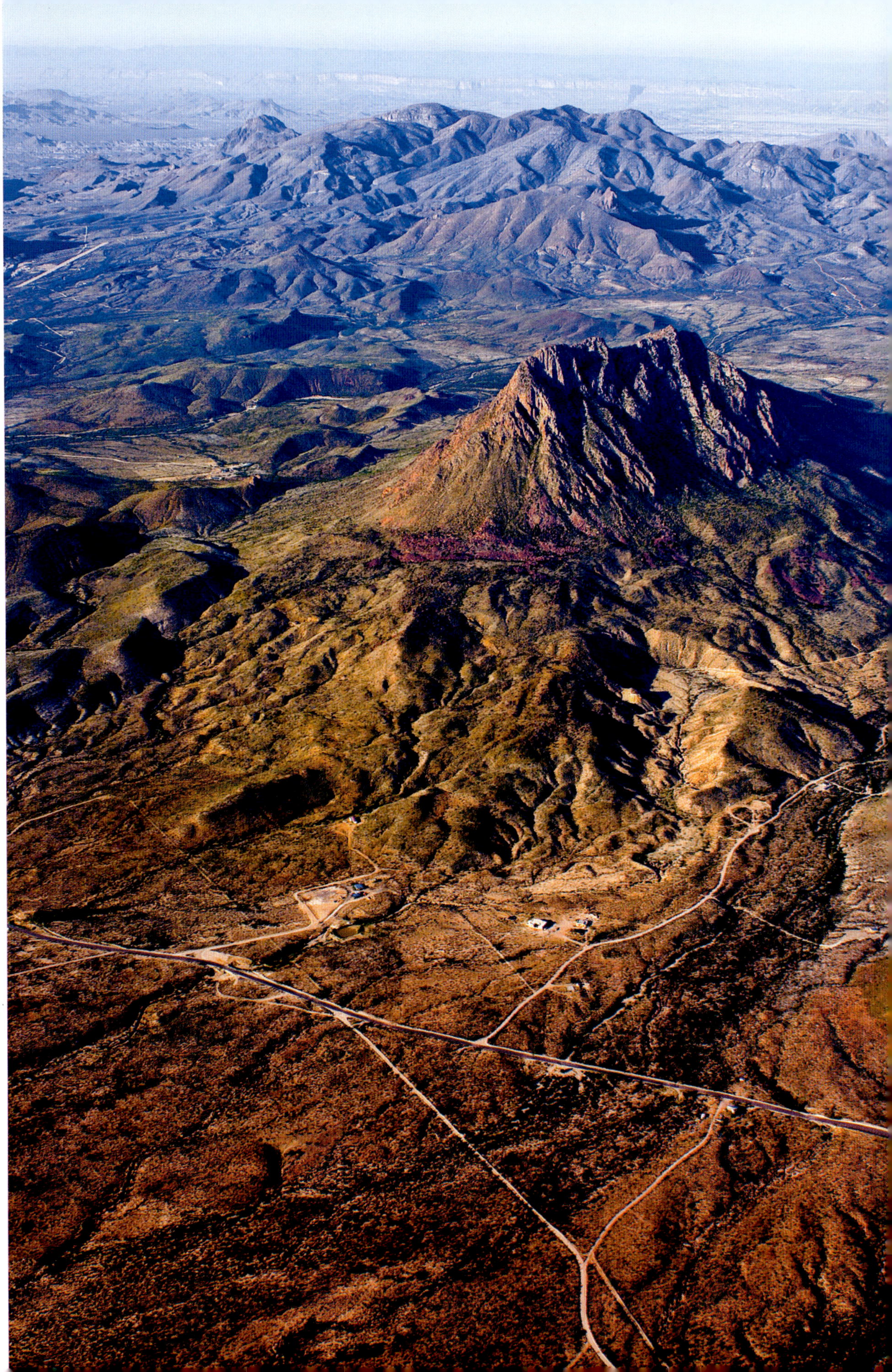

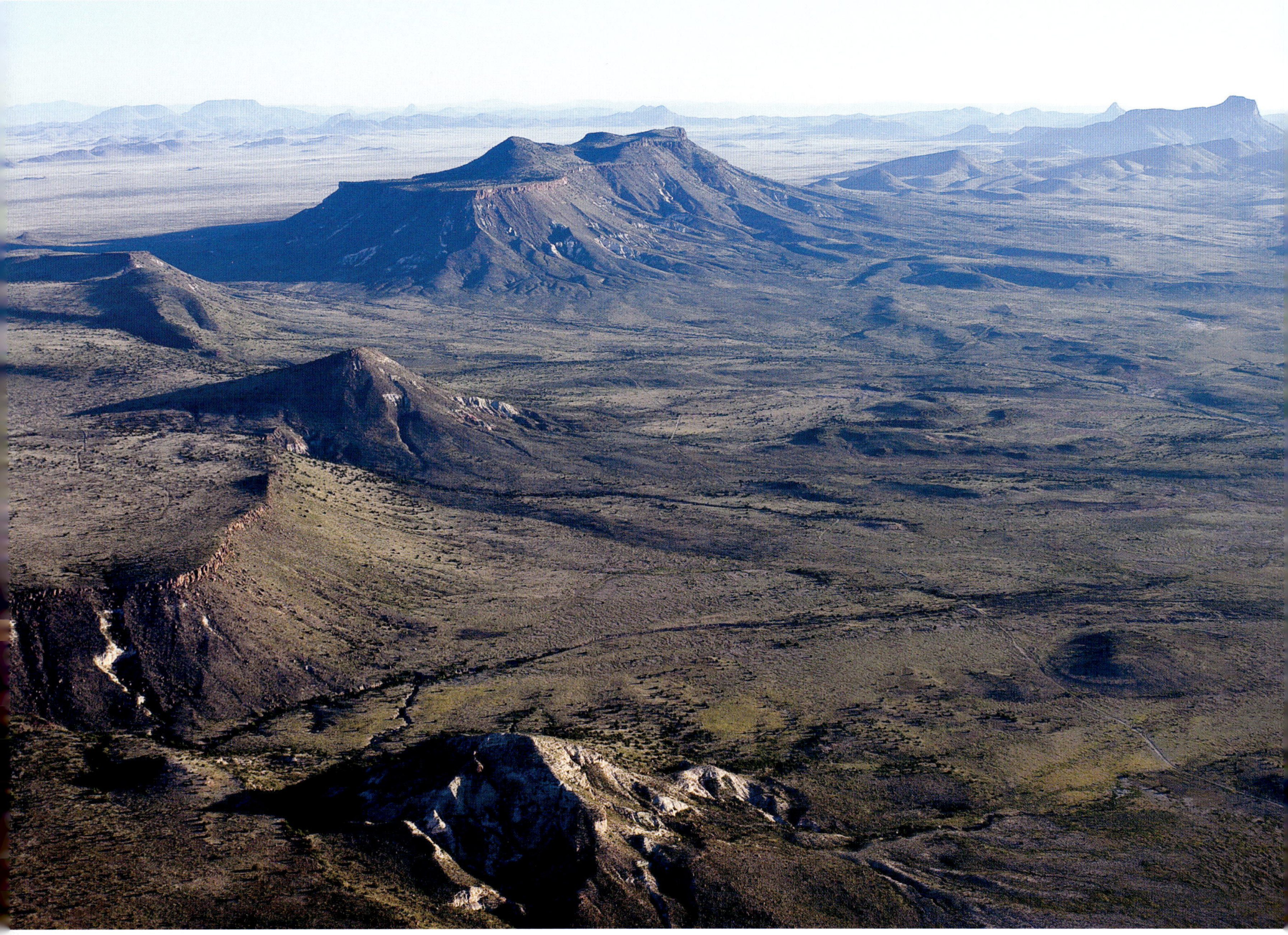

21. Marfa Lights, Backstage

West of Texas Highway 118

Cathedral Mountain is toward the north, with the OT Mesa in the foreground. Goat Mountain (upper center) forms the backdrop for people viewing the nighttime Marfa Lights, the mysterious phenomenon seen across the flats behind the mountain.

22. Kokernot Mesa in Morning Haze

South of Alpine

This is the view looking southeast down Dark Canyon toward Texas Highway 118 and Calamity Creek. The haze seen here caused problems during the aerial photography for this book. Most of the haze is pollution from coal-burning power plants in the United States. There are also two plants in the state of Coahuila, Mexico, that burn low-grade coal, putting 250,000 tons of sulfur dioxide per year into the atmosphere. Some coal-burning power plants located in Texas, by their age, are exempt from use of scrubbers on their smokestacks.

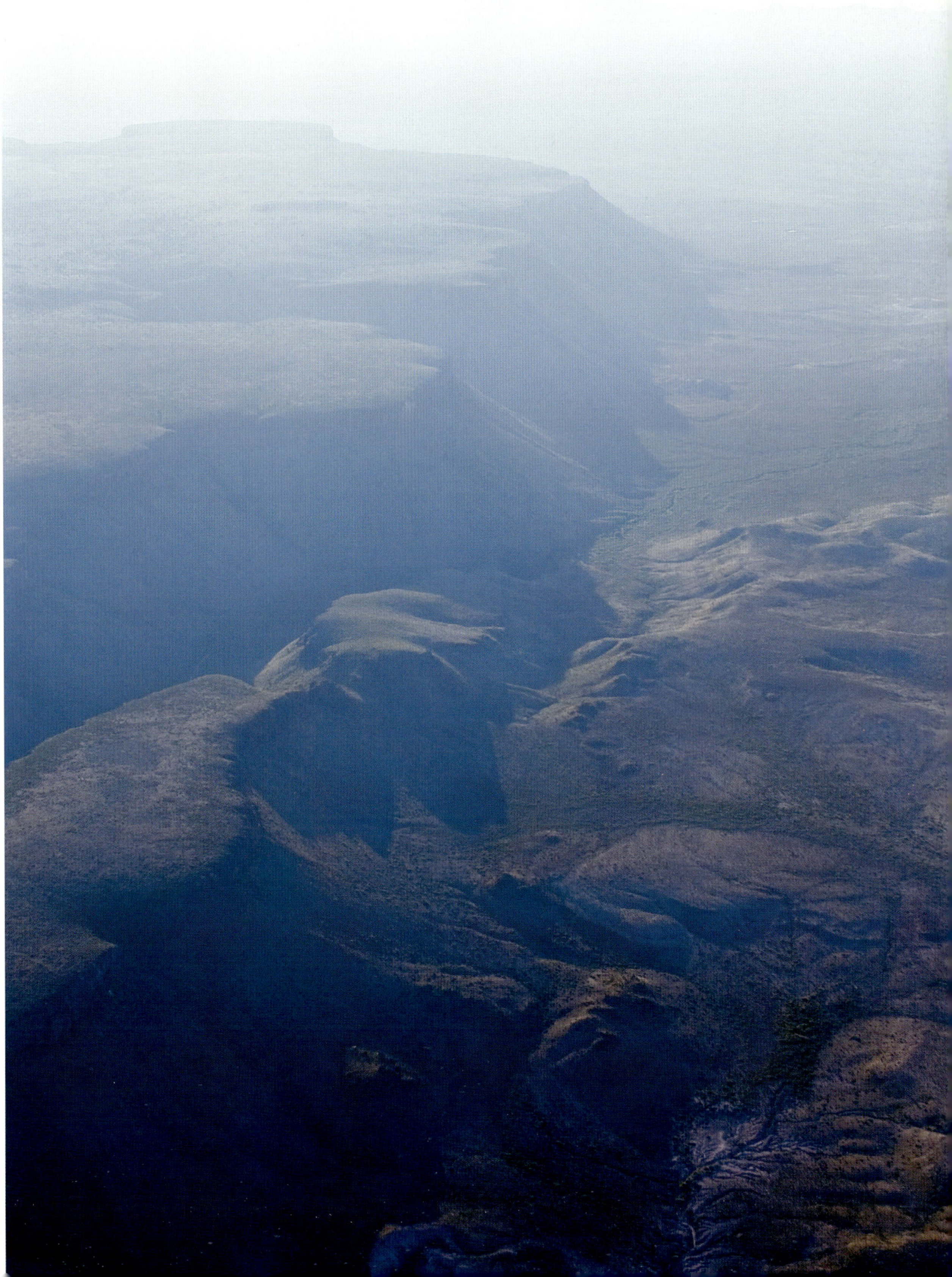

III. INTO THE HEART OF BIG BEND NATIONAL PARK

Exploring the Chisos Mountains

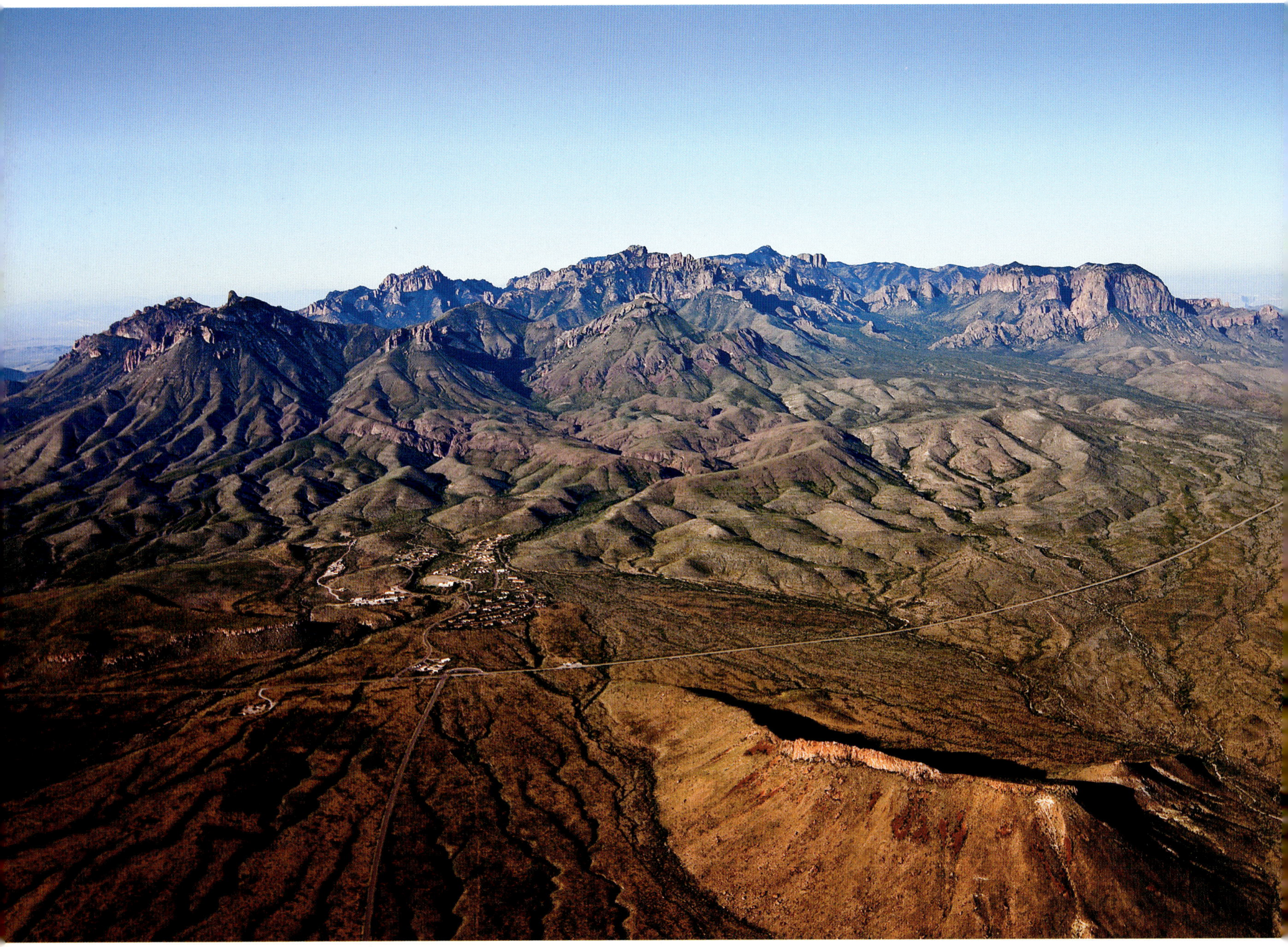

23. Chisos Mountains Vista

Big Bend National Park

Emerging from the surrounding arid flatlands, the Chisos are a mountain island in a sea of desert. Dynamic geologic forces of sedimentation, deformation, and volcanism formed the peaks such as Pummel Peak (left) and Pulliam Ridge (right). The Chisos Mountains form the centerpiece of Big Bend National Park, which covers 801,163 acres. I selected this as the first photograph of the Chisos Mountains in the book to help orient viewers to the aerial viewpoint. This approach is from the direction familiar to visitors, with Panther Junction visible in the foreground.

24. Panther Junction

Big Bend National Park

Panther Junction is the headquarters for Big Bend National Park. In the foreground Lone Mountain, with its erosion-resistant igneous rock cap, dominates the view. In the background, just to the right of center, is Pummel Peak, shaped like the pummel (or pommel) of a saddle. This view appealed to my photographer's eye because of the composition in which Lone Mountain points to the peak in the distance, leading our eyes like an arrow on a road sign.

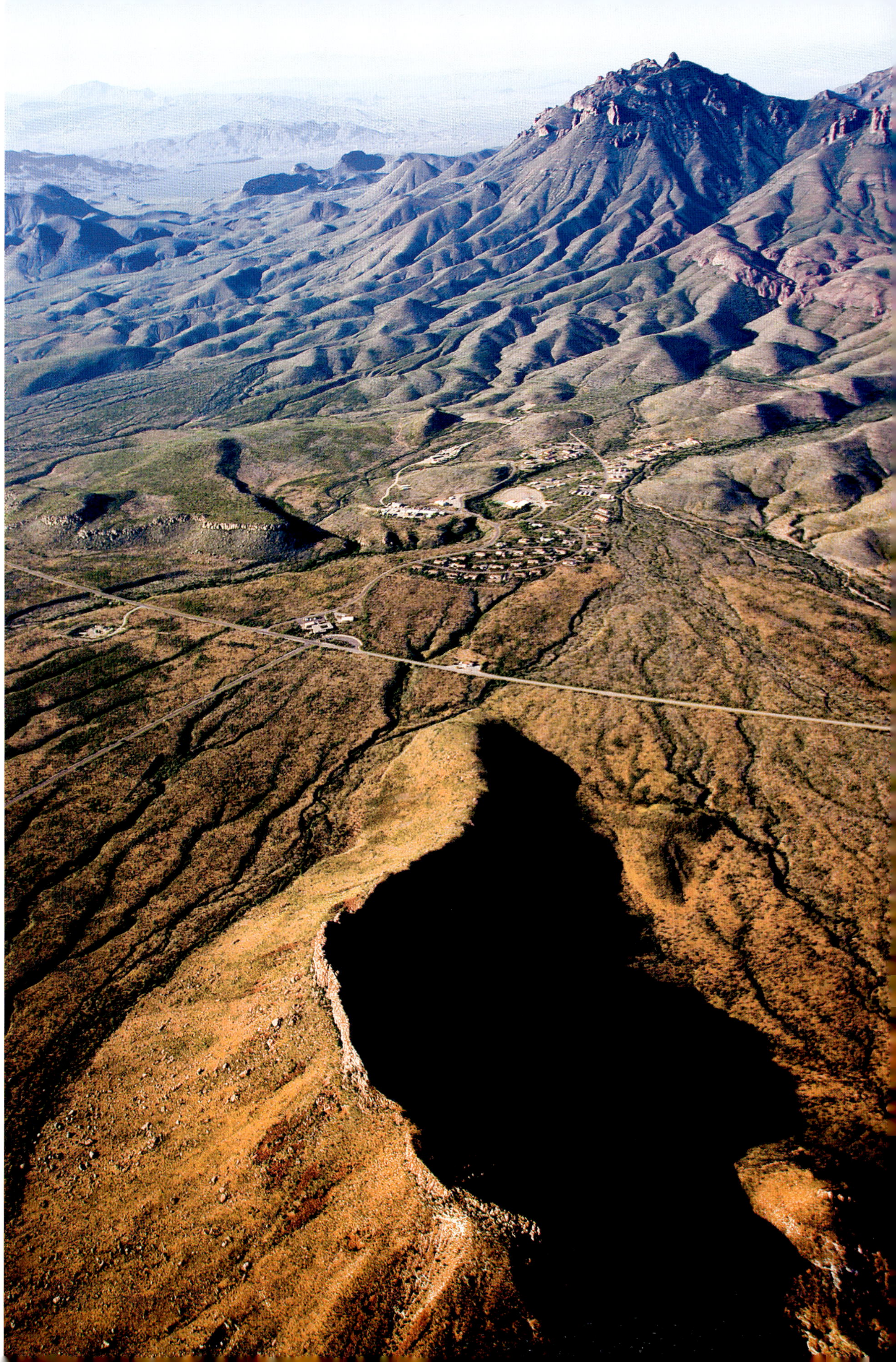

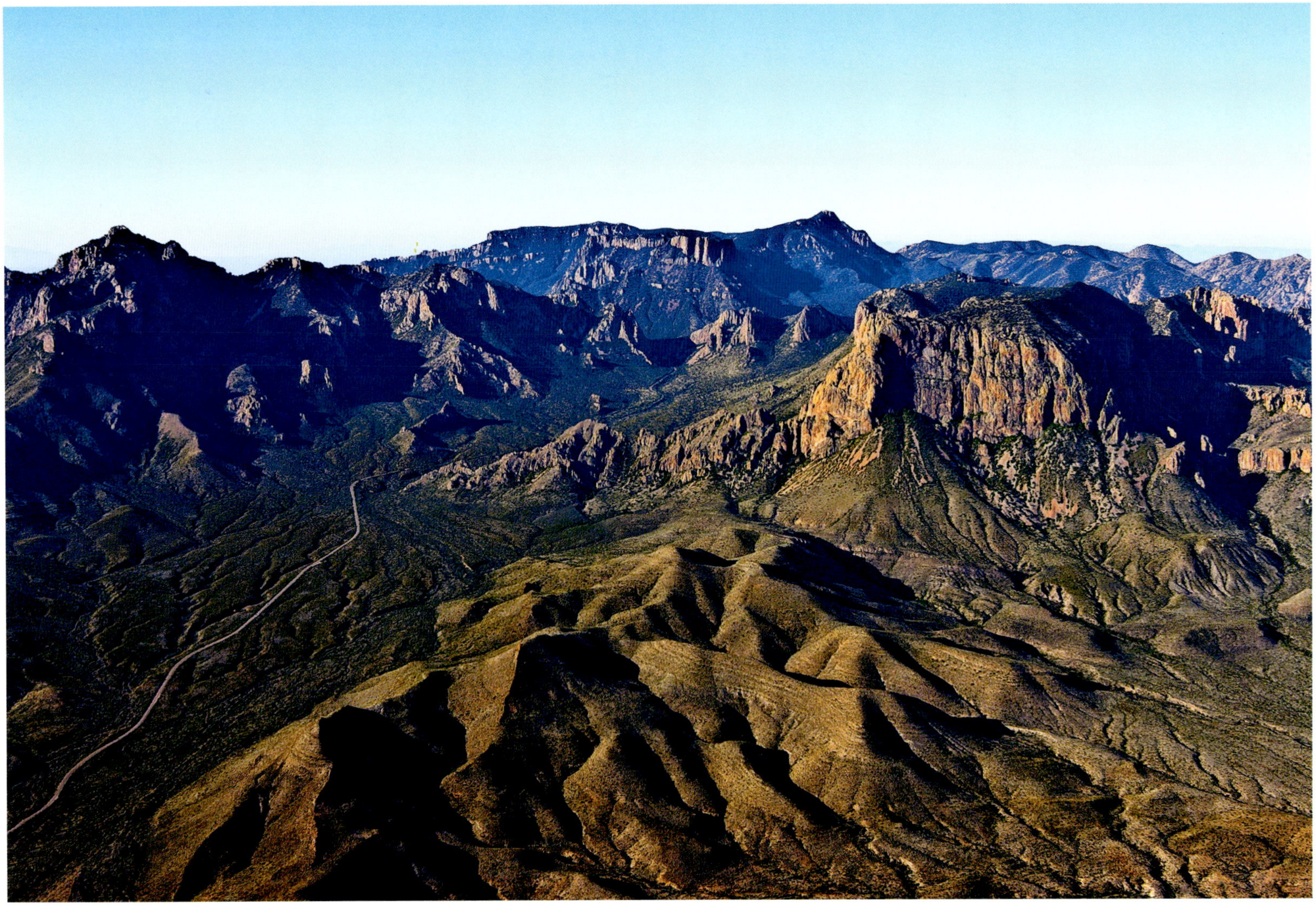

25. Green Gulch

Chisos Mountains

Basin Road, the main access into the Chisos Basin, winds uphill through the high desert grassland and woodlands of Green Gulch to Panther Pass. On the right is the sheer igneous rock face of Pulliam Ridge. In the distance at left is Lost Mine Peak, the supposed location of a lost secret Spanish gold mine. The Civil Conservation Corps built the historic road in the 1930s.

26. Northwest Side of Pulliam Ridge

Chisos Mountains

Pulliam Ridge was formed when magma filled a cavity in the earth. When the magma cooled and the surrounding material was washed away by erosion, the ridge was revealed. This formation is named after Bill Pulliam, a legendary Big Bend rancher who cut pine trees from the ridge and lowered them by cable to build a log cabin for his wife.

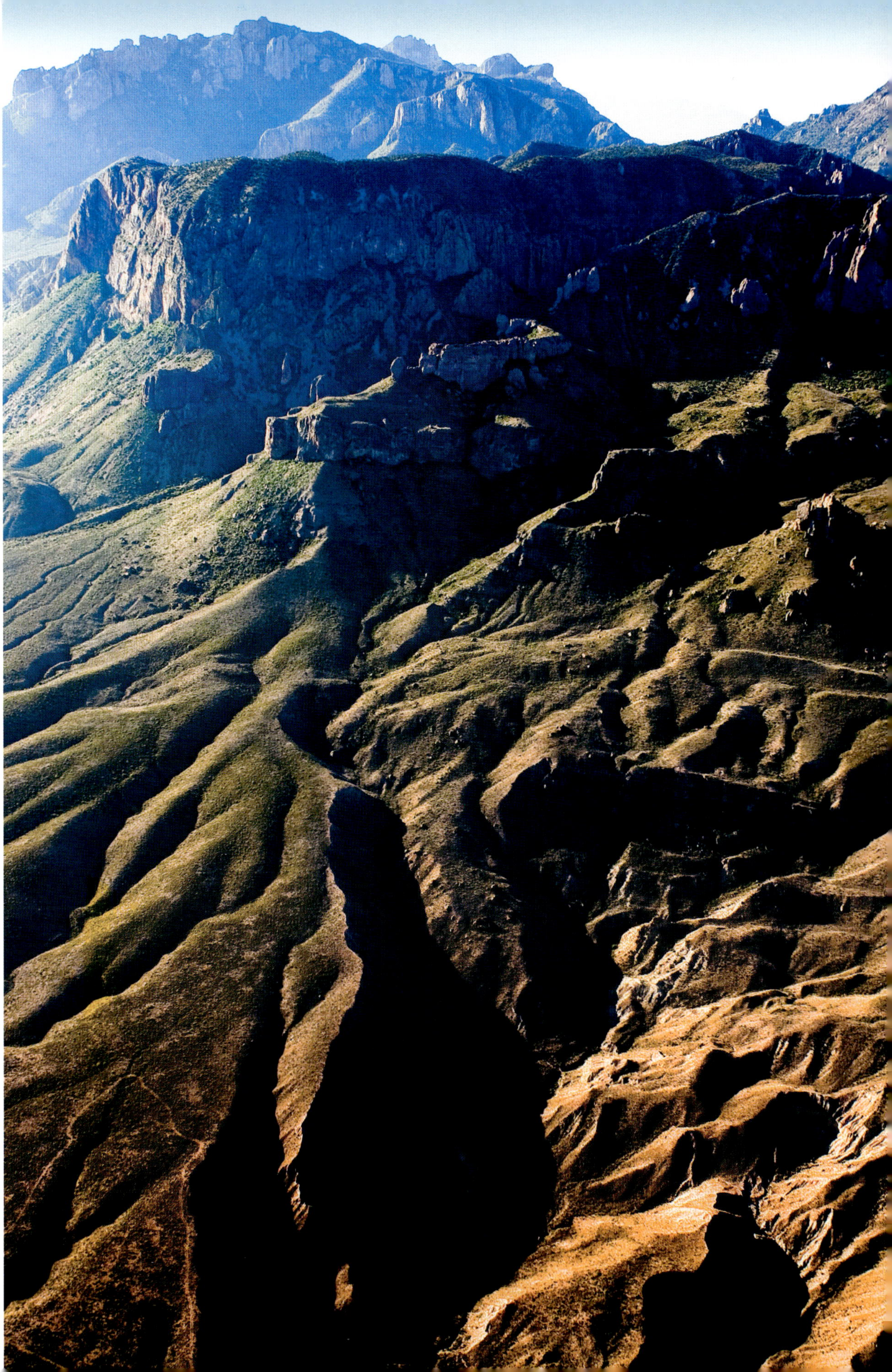

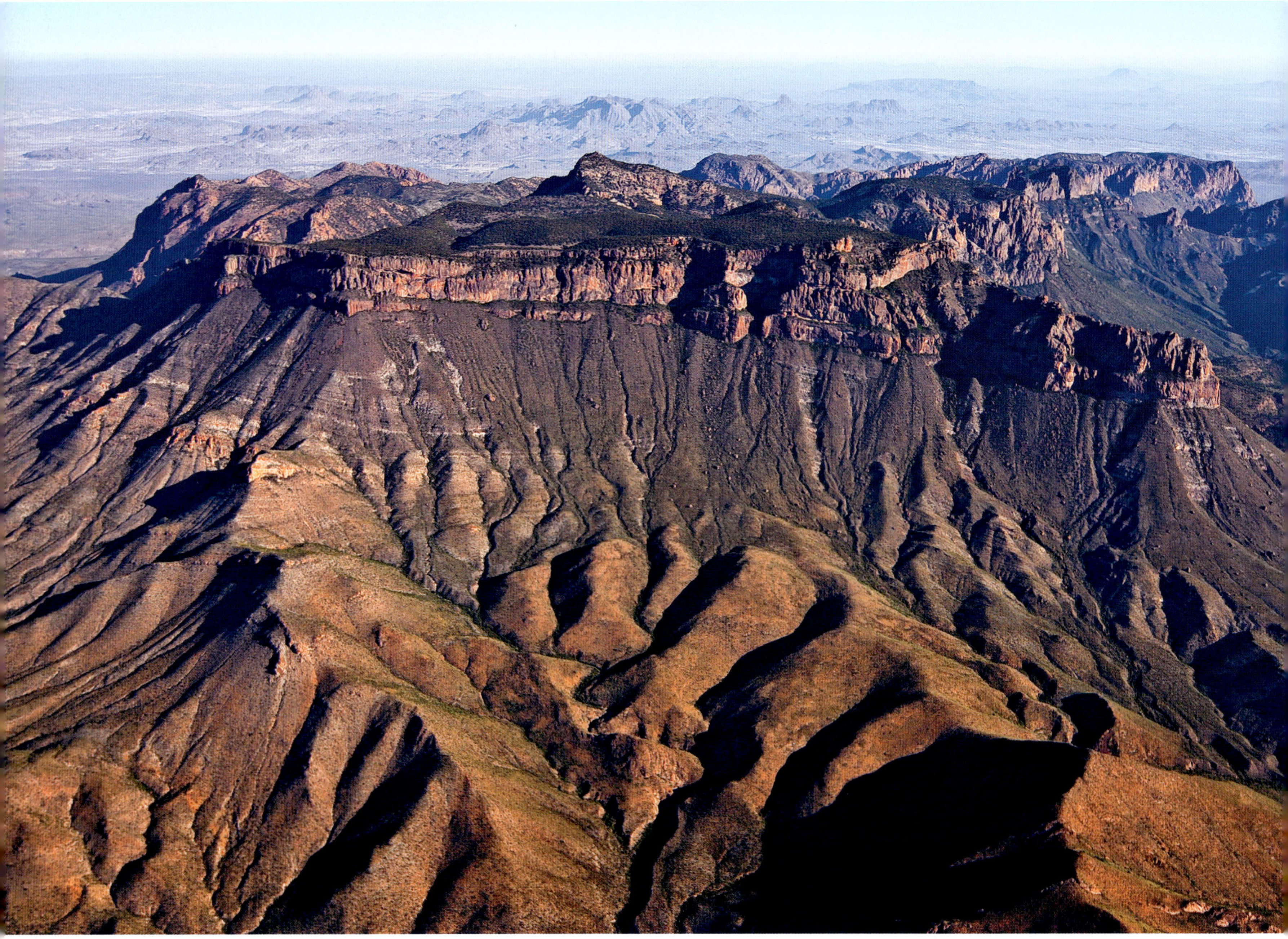

27. South Rim and Emory Peak

Chisos Mountains

The hike from the Chisos Basin trailhead to the South Rim is widely considered the most strenuous and spectacular day hike in Big Bend. Emory Peak (center) rises an additional 1,000 feet above the rim. The twelve-mile loop takes at least eight hours to complete, with a vertical ascent of 2,000 feet, and hikers have reported mountain lions along the trail. The South Rim trail is closed at certain times of the year when peregrine falcons nest on the cliffs.

28. Lower Juniper Spring

Chisos Mountains

Townsend Point (left) and Upper Juniper Canyon (right). According to Native American legends, ancient spirits inhabit the Chisos Mountains. Hikers have reported unexplainable glowing lights on the mountainsides. This is one of my favorite images of the series due to the beautiful palisades on the upper rim and the sculptural shapes of the lower foothills that seem to guard the mountains above.

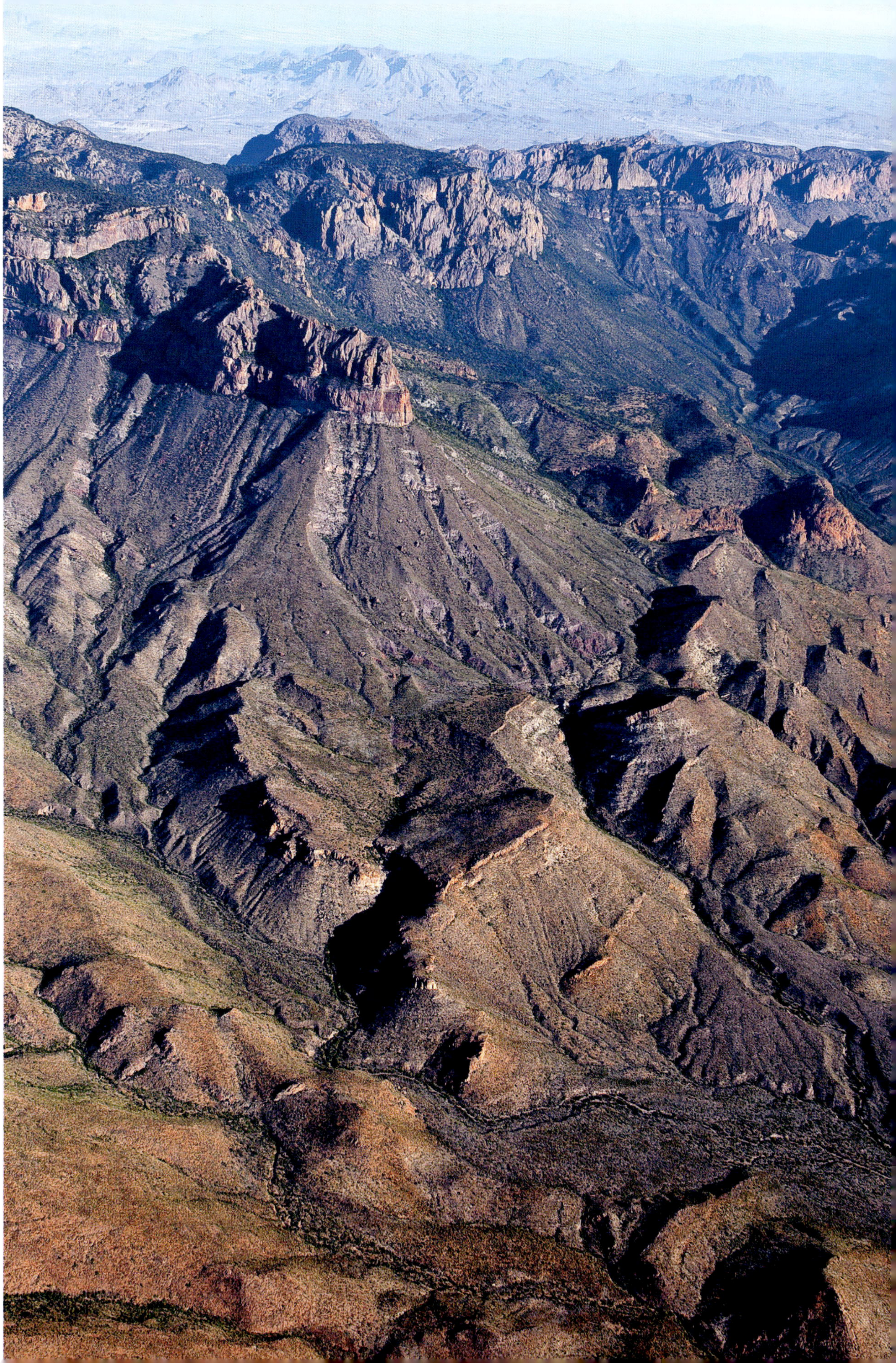

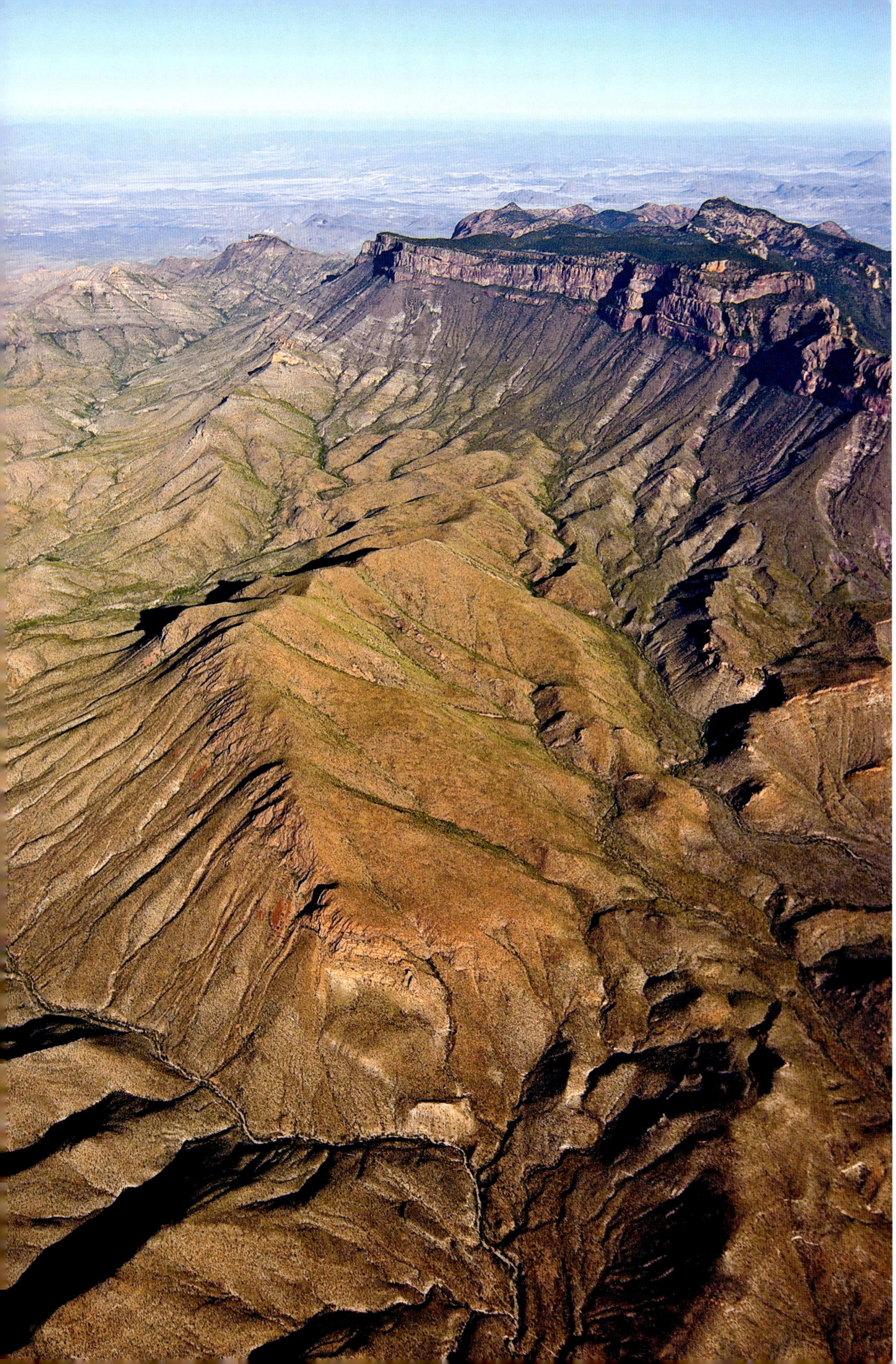

29. Fresno Creek

Sierra Quemada

The mountains at the base of the Chisos are called Sierra Quemada, which means "burnt mountains" in Spanish. The hills are covered with highly flammable grass, which is periodically ignited by lighting strikes, then rejuvenated after rains.

30. Approach to Pine Canyon

Chisos Mountains

The verdant valley of Pine Canyon leads upward toward the grassy saddle of Smuggler's Gap (top center), which forms a natural pass through the mountains. Pine Canyon was also known as Wade Canyon, named for Lloyd Wade, who hauled the first bathtub into the mountains for his cabin there. To the right of Smuggler's Gap is Panther Peak. Mountain lions still roam the Chisos Mountains, and there are over 150 sightings of the big cats each year in Big Bend National Park. There have been four attacks on humans since 1984. These big predators naturally regulate the deer and javelina populations.

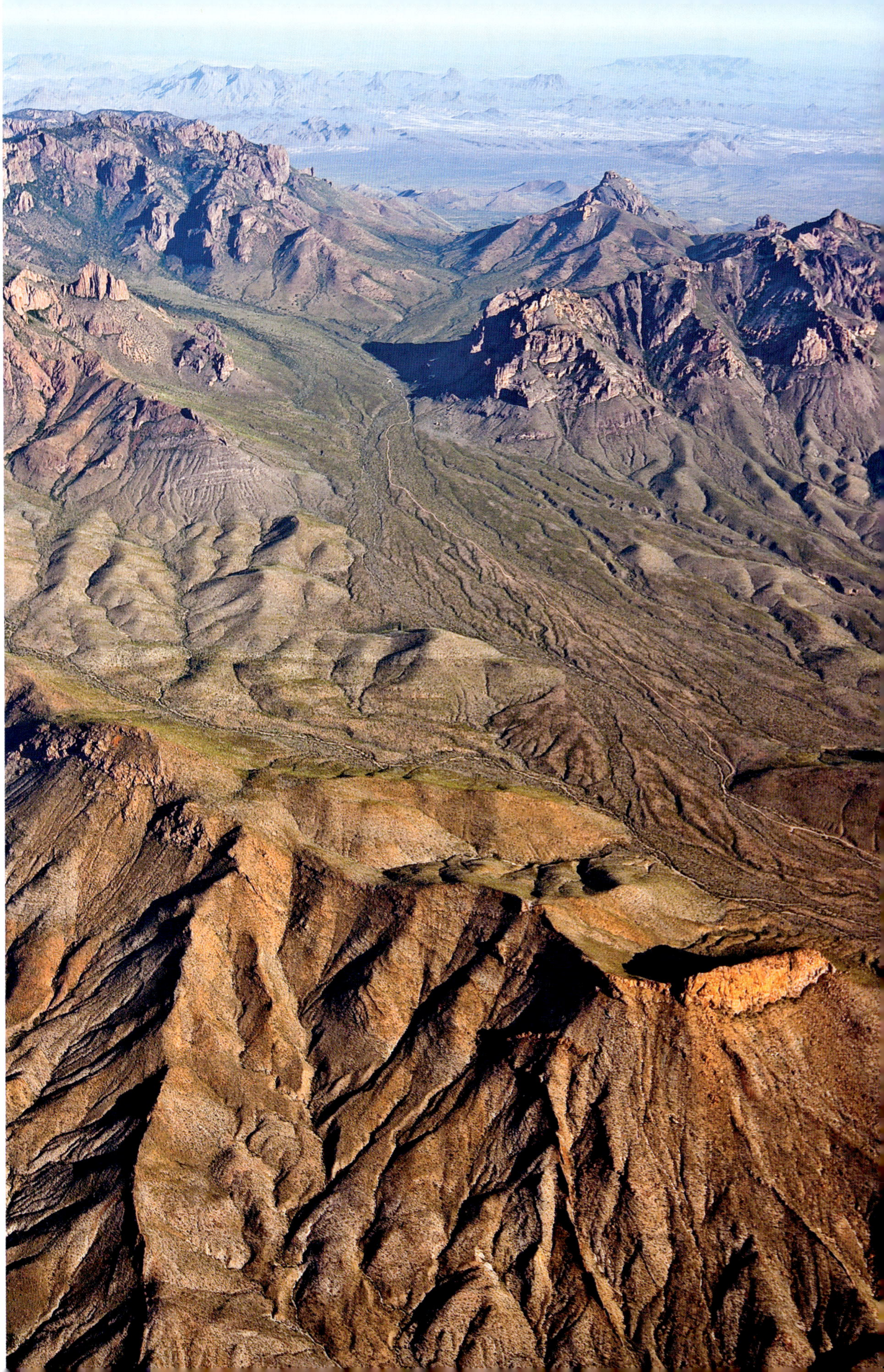

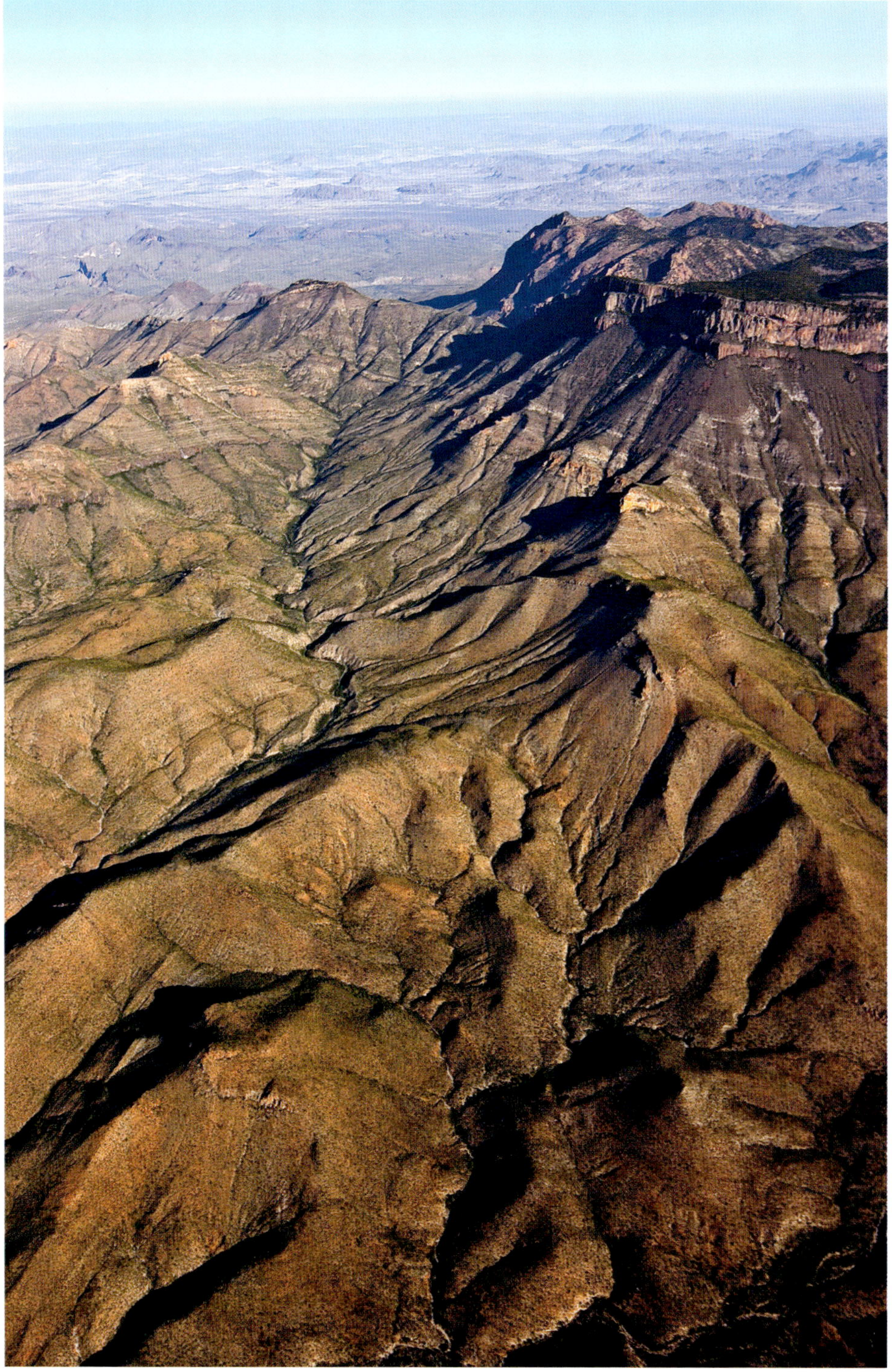

31. Over the Dodson Trail

Sierra Quemada

Just below the South Rim of the Chisos Mountains (upper right), Fresno Creek flows twenty-one miles southeast to the Rio Grande. Dodson Spring (lower right) was the water source for Harve Dodson's ranch, which was located nearby. The cabin ruins are still visible along the Dodson Trail, which he used to transport supplies through the mountains for his family. Recreational hikers now utilize the trail to explore the Sierra Quemada.

32. Crown Mountain

Chisos Mountains

Soaring to 7,010 feet in elevation, the multiple spires of Crown Mountain rise above Juniper Canyon (left). In the summer the cooler high altitudes attract the Mexican black bear, an endangered species in Texas that roams the Chisos. One of the favorite foods of bears in Big Bend National Park is the succulent base of the Texas sotol plant. The bears also feast on wild honey as well as berries and nuts from juniper, pinon pine, and madrone trees.

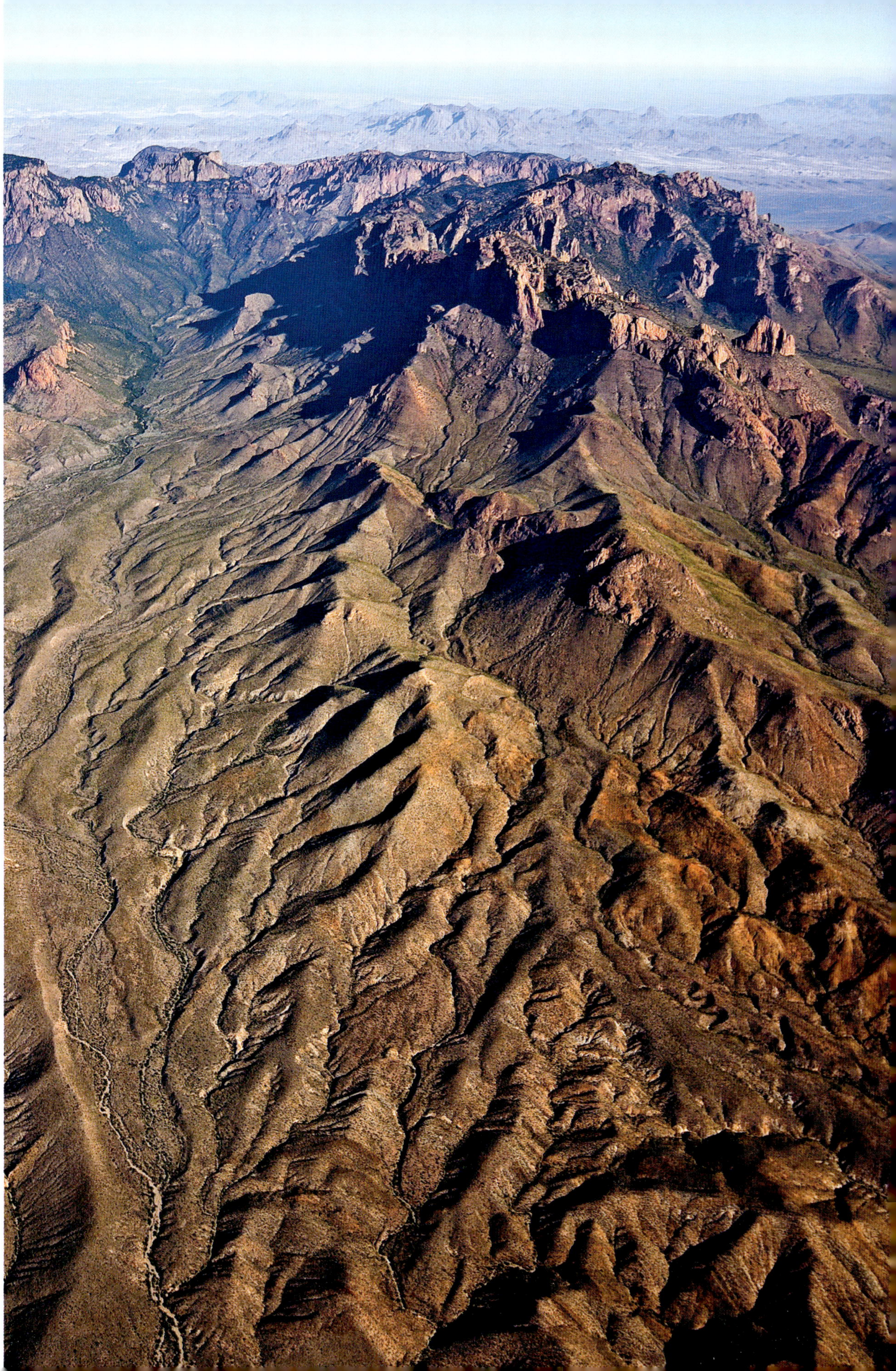

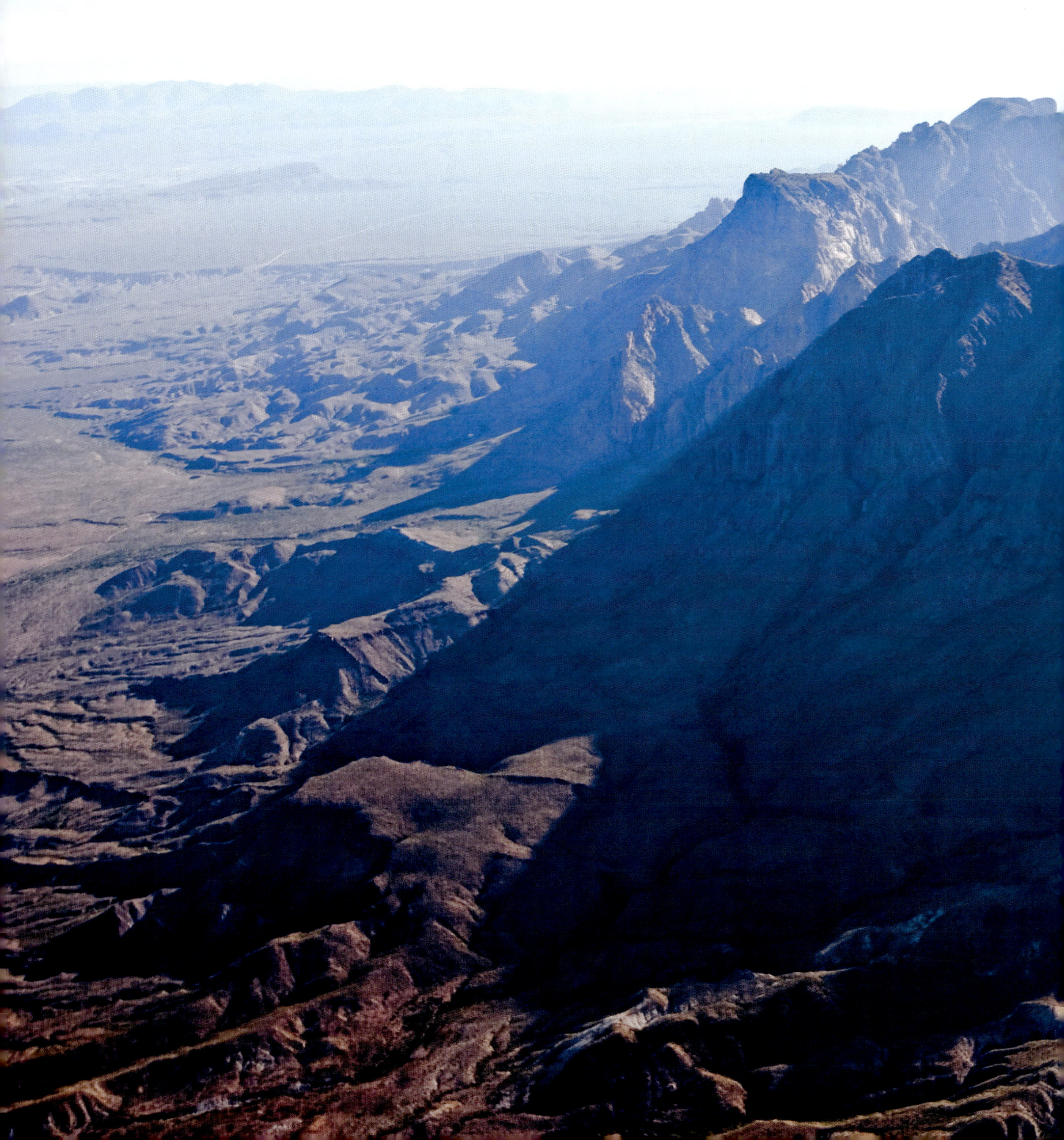

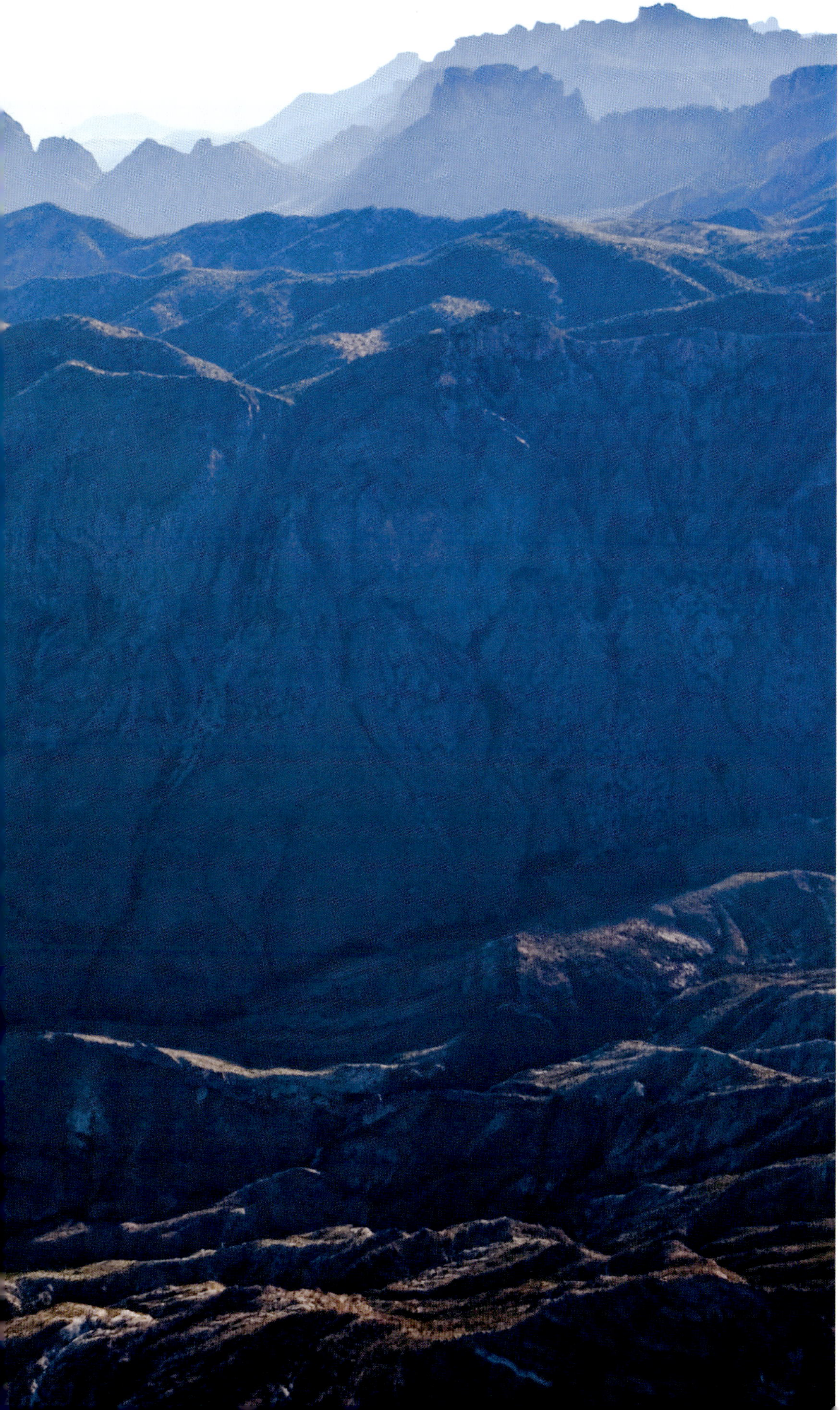

33. Across the Window

Chisos Mountains

Sunlight pours through the Window, which is the notch between Ward Mountain (foreground) and Vernon Bailey Peak on the western side of the Chisos. Along the right edge of the photograph (bottom to top) are Blue Creek drainage, Ward Mountain, Casa Grande, and Lost Mine Peak. As we turned the southwest corner of the Chisos range, I saw the eastern sunrays as they raked across the peaks and illuminated the valleys. The light created an image of the cathedral-like grandeur of these mountains. I remembered the Native American legend that the oldest spirits in the earth live here, the sacred place where braves made their vision quests. High above these mysterious peaks, I captured an image that communicates the airy, dreamy quality of Big Bend and the euphoria of the flight.

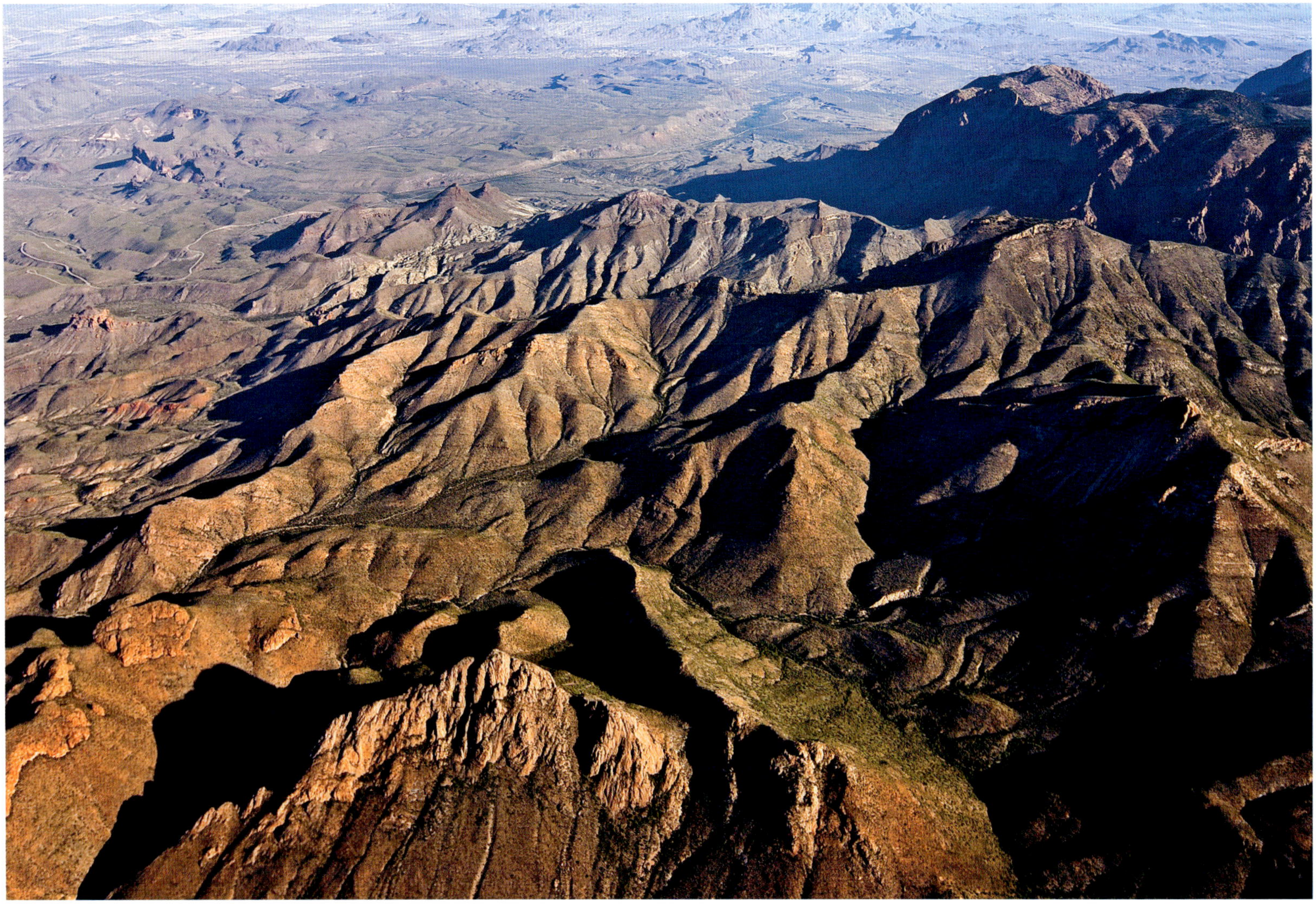

34. Looking toward Ross Maxwell Scenic Drive

Sierra Quemada and Chisos Mountains

From upper right to lower left are Cattail Canyon, Ward Mountain, Blue Creek, and Knife Edge. In the distance at the far left (top to bottom) are Burro Mesa, Sotol Vista Overlook, and Carousel. The flatlands are covered with sotol, a plant that throughout the history of Big Bend has provided desert dwellers with material for shelters, rope, baskets, and even liquor.

35. Beyond Blue Creek to Burro Mesa

Sierra Quemada

Above the pointed peak are Javelina Wash, Ross Maxwell Scenic Drive, and Burro Mesa. Burro Mesa is named for the wild burros that used to roam there. In the middle ground is Blue Creek, running from right to left amid the Sierra Quemada. In the summer of 2011 the National Park Service managed a 2,000-acre wildland fire in this area. The fire cleared debris and undergrowth as natural fires have for ages, allowing the land to renew itself.

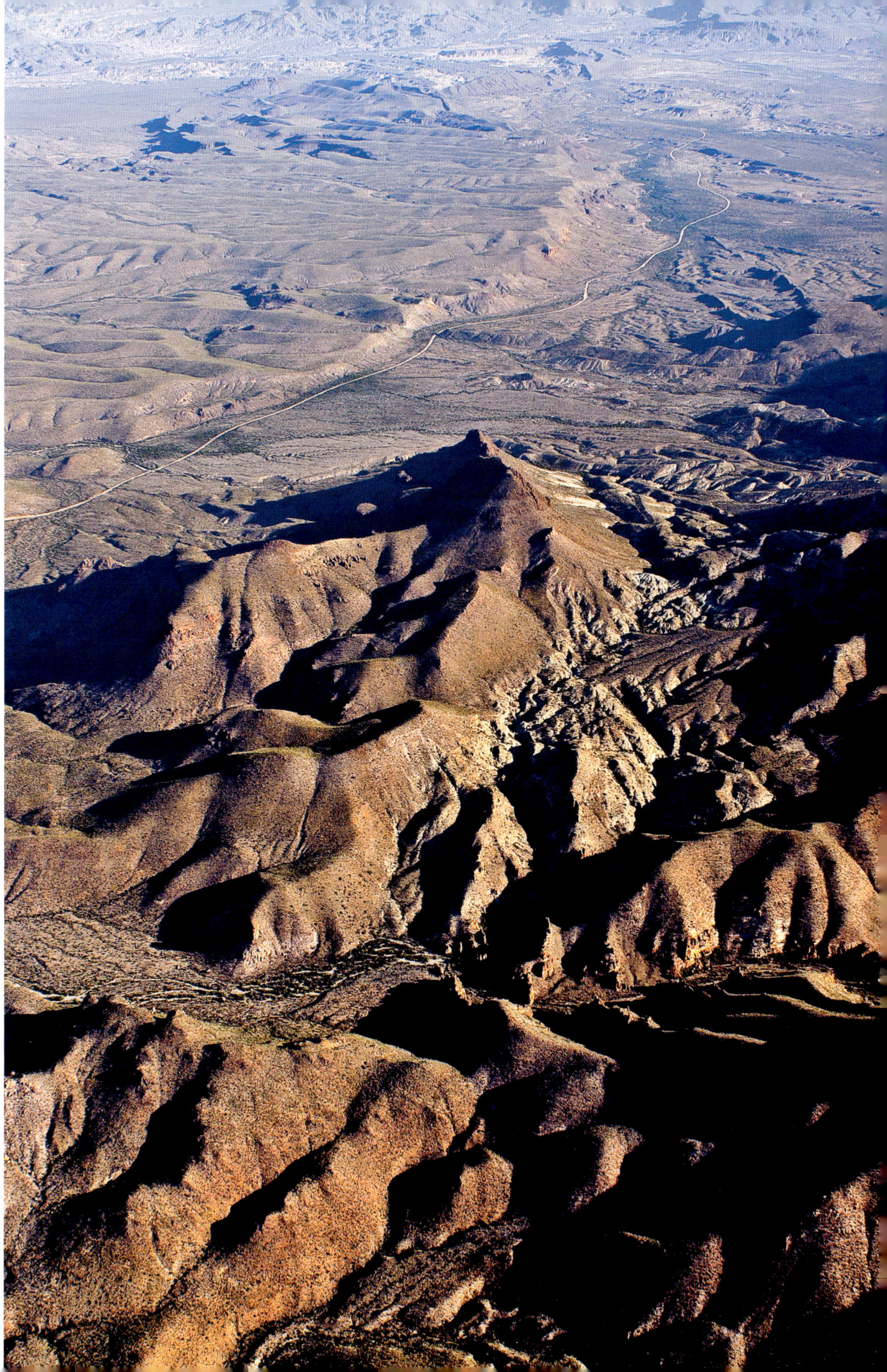

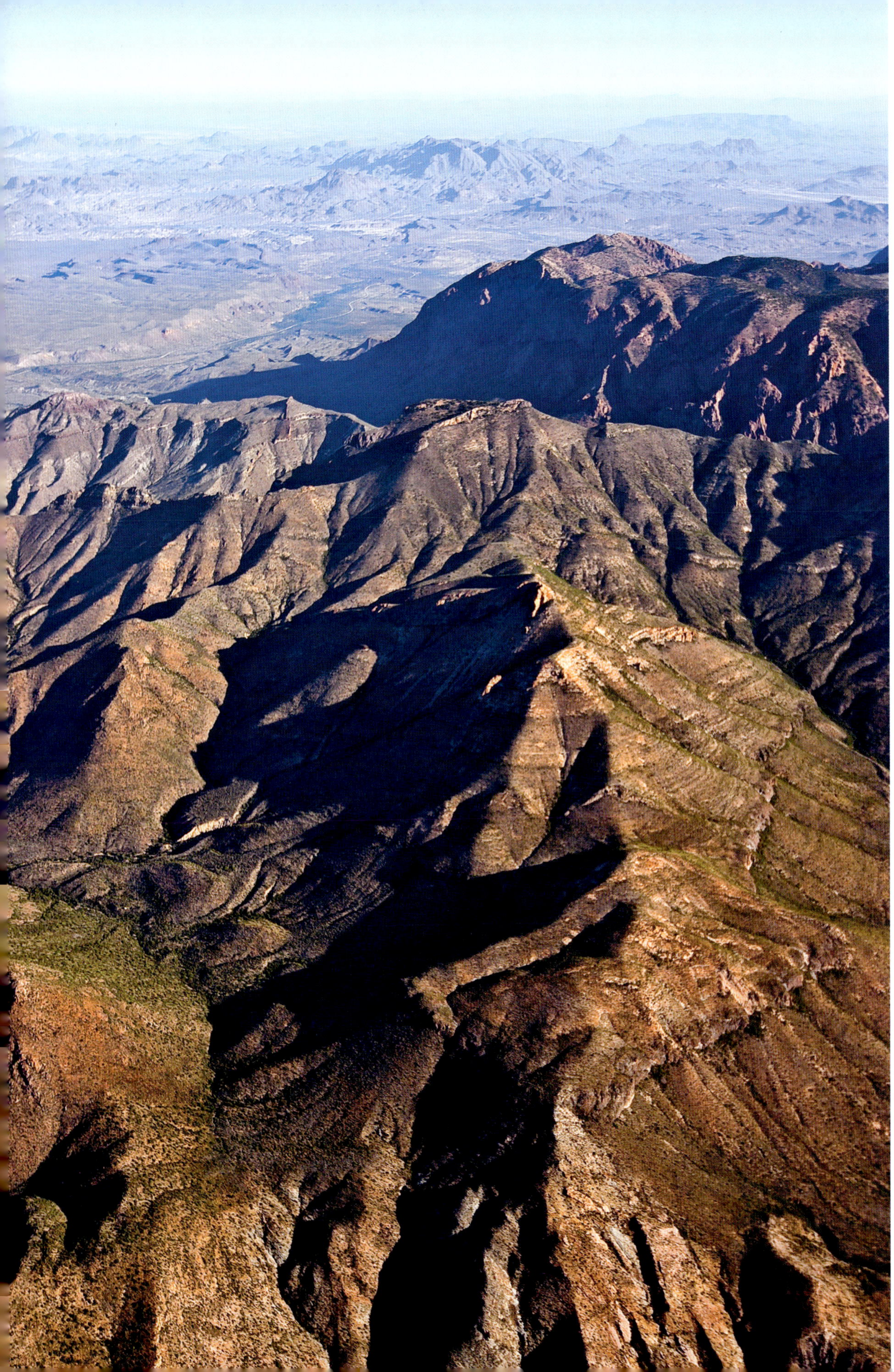

36. Ward Mountain

Chisos Mountains and Sierra Quemada

Ward Mountain is named for Johnnie Ward, a cowboy who camped here when he worked on the G4 Ranch. The G4 was one of the largest Trans-Pecos ranches in the late 1800s. The green band of vegetation just to the left of the mountain is watered by Cottonwood Creek. The subtle, almost painterly markings on the hillsides piqued my interest from this camera angle.

37. The Divide of Smoky Creek and Blue Creek

Sierra Quemada

These mountains are located between Smoky Creek and Blue Creek, just north of the Dodson Trail. Near here are the one-room rock cabin and ruins of the corral and cistern built by Homer Wilson. Just above this view is a 5,360-foot peak. I could see this view of the south face of the mountains appearing as we flew east. I had to wait for us to get the aircraft into just the right position for this angle. I readied my camera, waiting for the moment when this combination of mountain ridges would fill the frame. When I pressed the shutter button, I knew that this would be a special photograph. The fact that it was taken with a normal and not a telephoto lens, not "zoomed-in," attests to how close we flew to the mountains; they seemed to be just off our left wing. This particular image stood out for the way it captures the pure visual drama of these entrancing mountains.

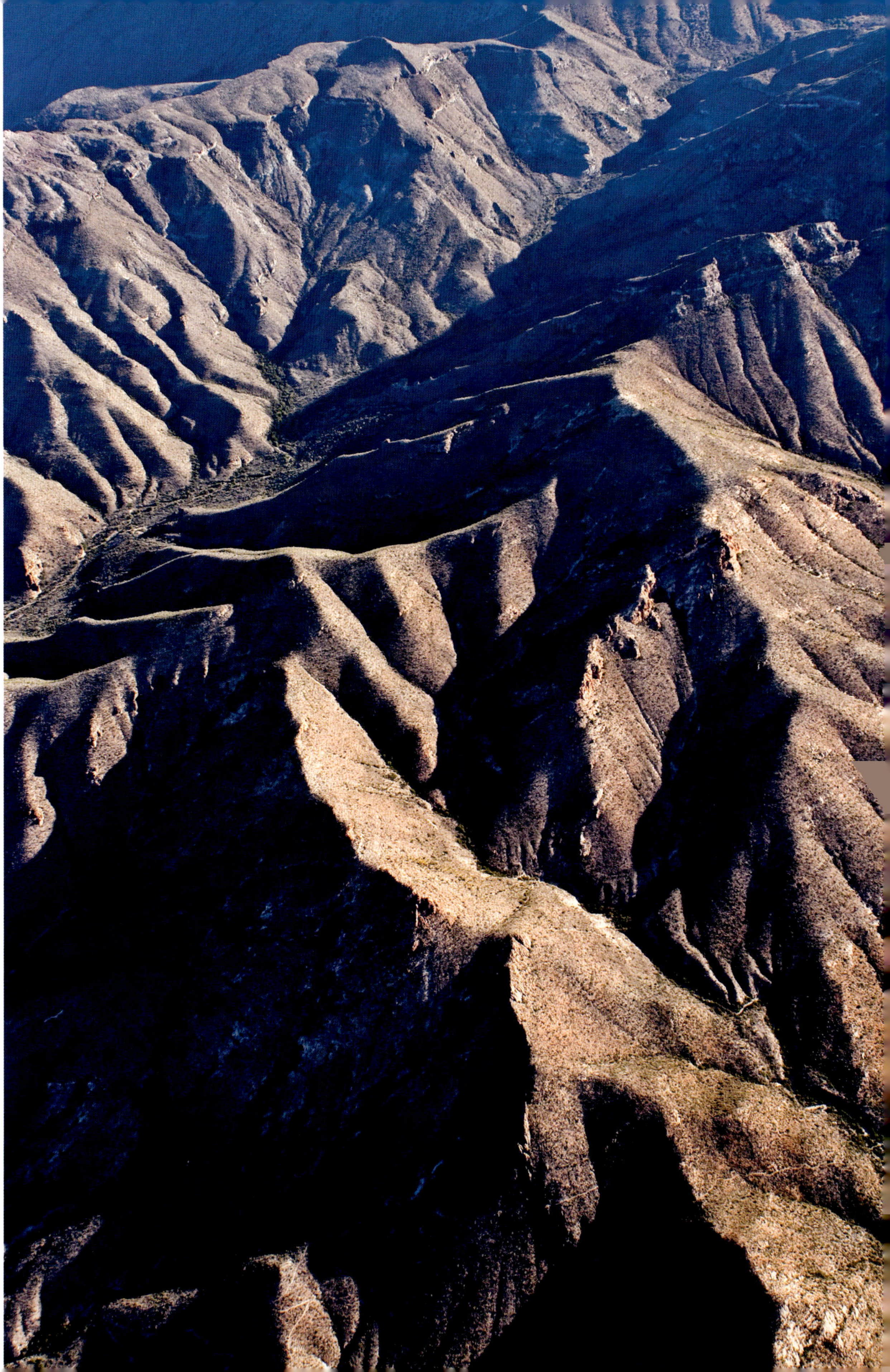

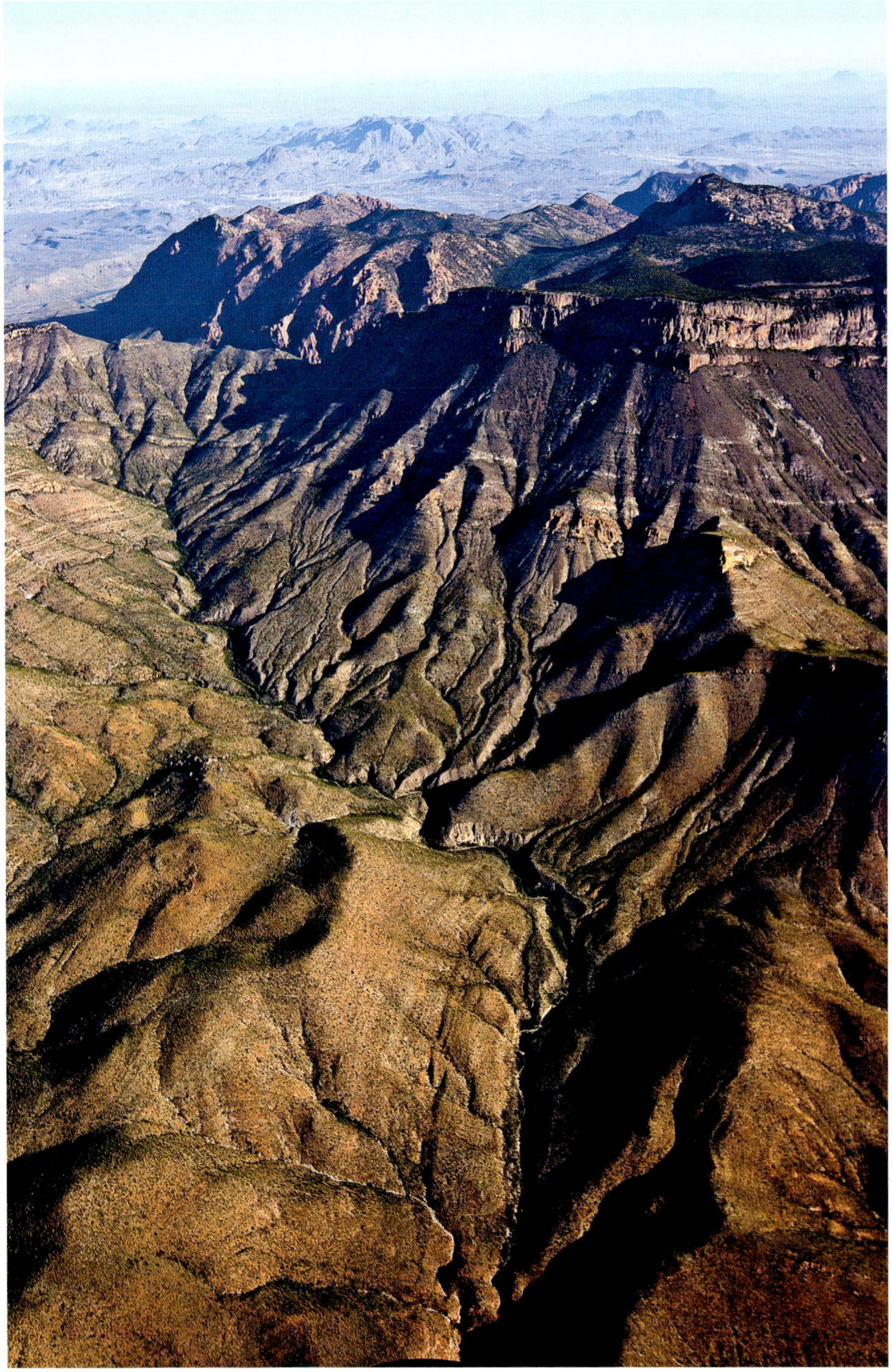

38. Upper Fresno Creek

Chisos Mountains

Just below Emory Peak and the South Rim flows Fresno Creek. *Fresno* is the Spanish word for the ash tree, whose durable wood was used by settlers to make the axles and brakes of wagons. I made this image to show the contrast in landscape textures displayed here by the exposed rugged rocks of the South Rim and the smooth stippled slopes of the Sierra Quemada.

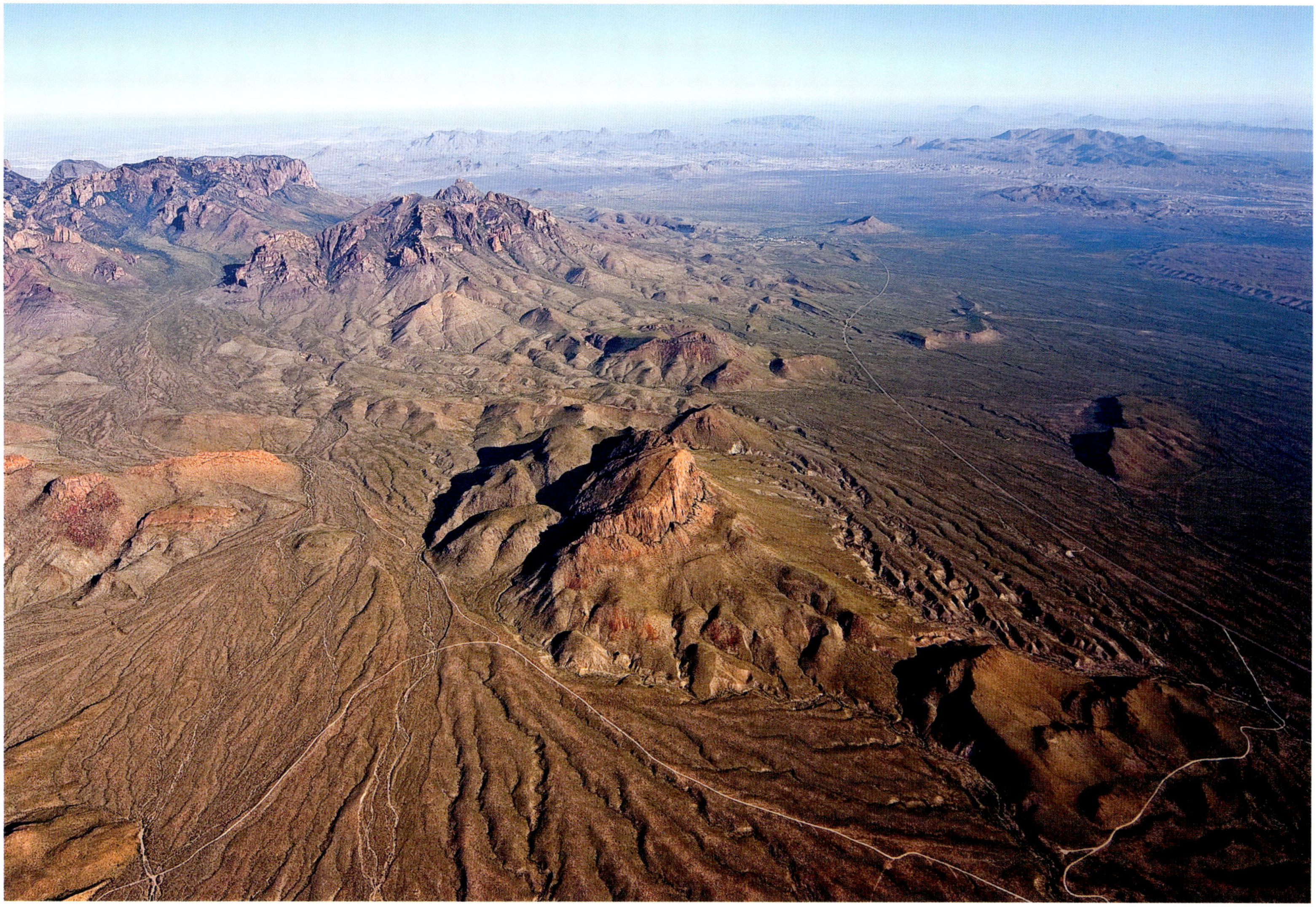

39. Chisos Mountains Panorama

Big Bend National Park

Lower Pine Canyon and Nugent Mountain (center) dominate this view of the sky island of the Chisos Mountains rising from the Chihuahuan Desert. Here the mountains and desert overwhelm the marks of humankind. In the foreground Glenn Spring Road threads a sinewy path south of Nugent Mountain. At right (bottom to top) are Dugout Draw, Estufa Canyon, the Estufa Hills, the Grapevine Hills, and the Rosillos Mountains. This camera angle reveals how the east end of the mountains drops away dramatically to the seeming infinity of the desert, and captures the sense of awe that Big Bend inspires in Texans and visitors from around the world.

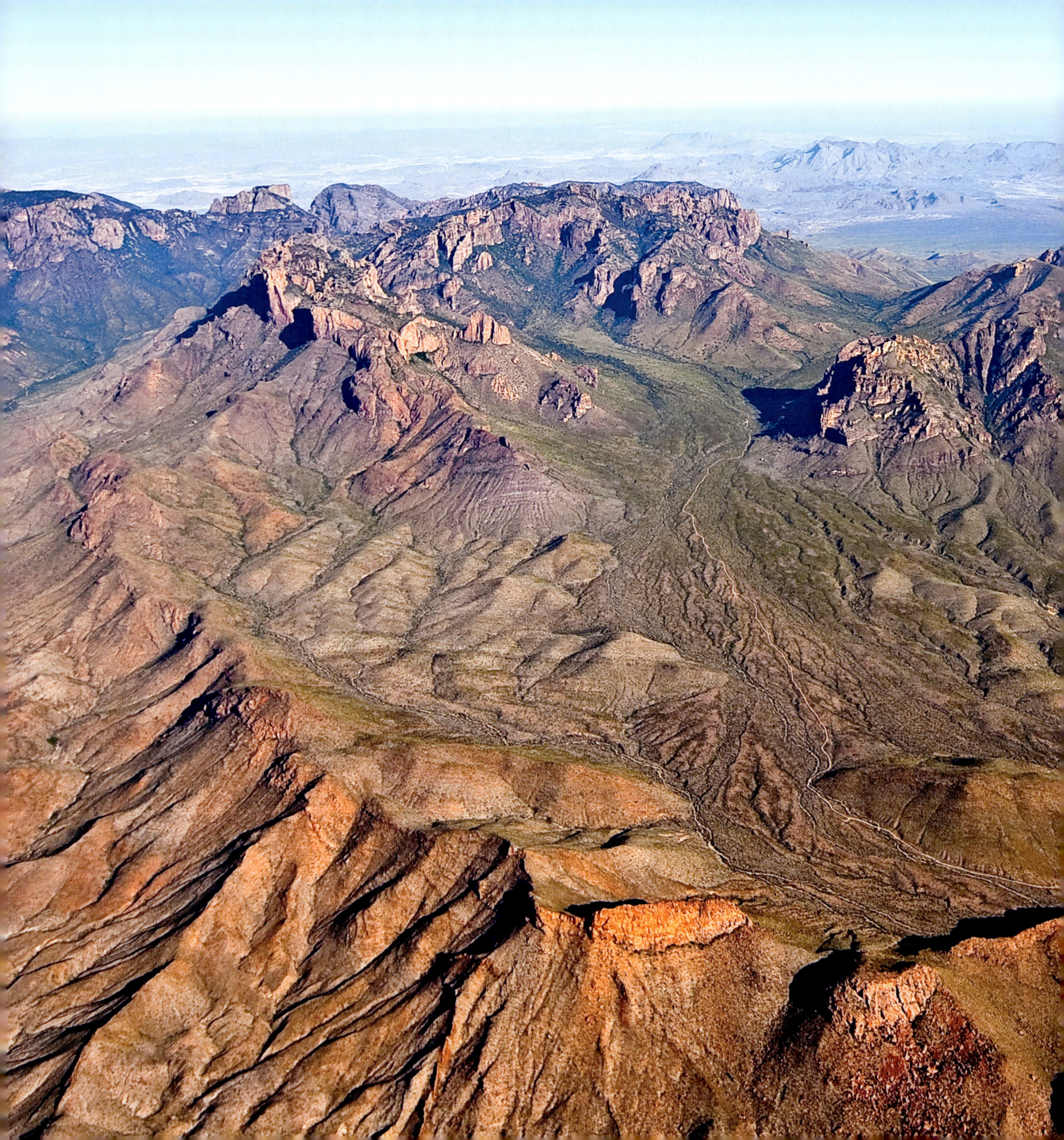

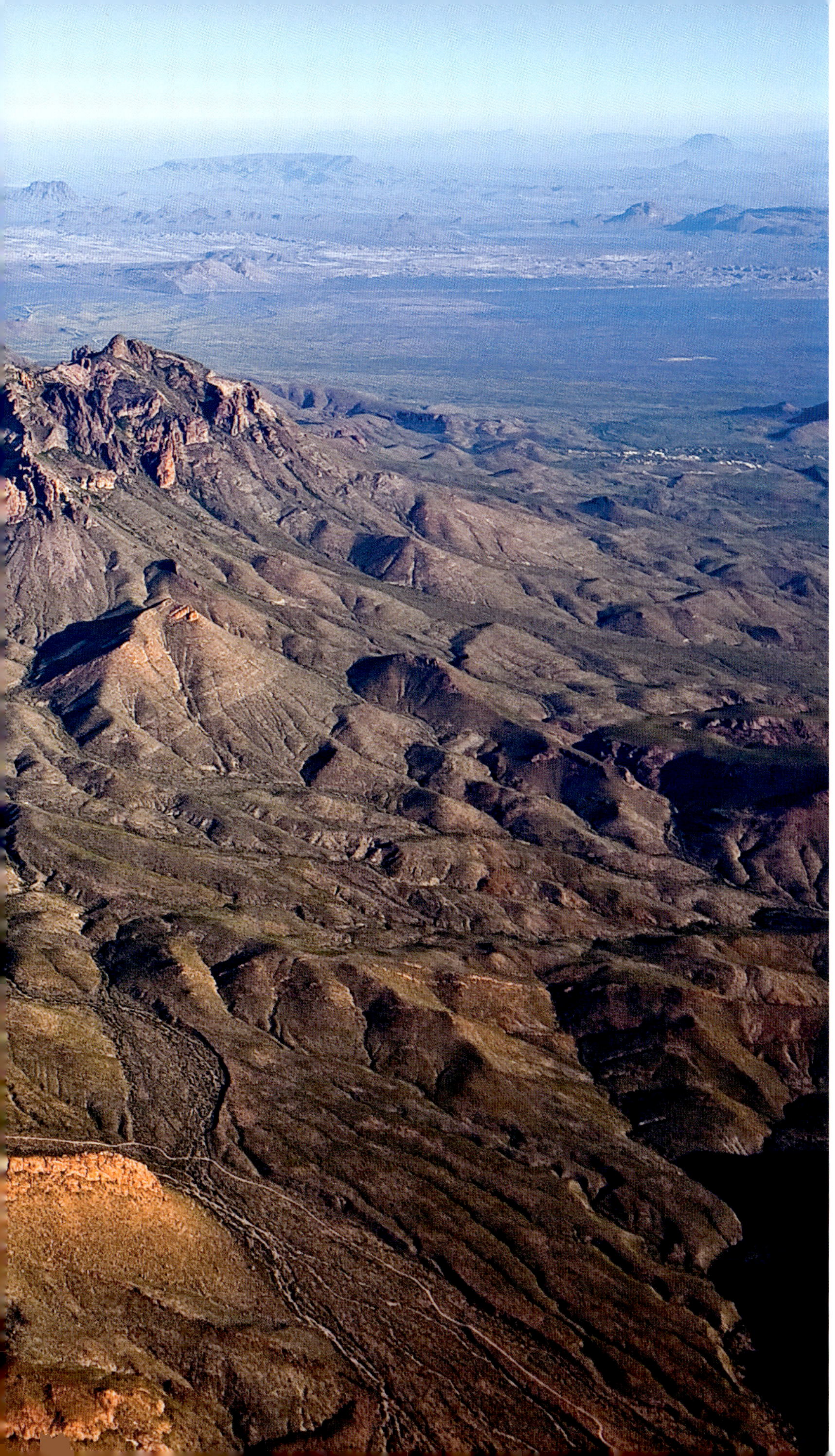

40. Pine Canyon

Chisos Mountains

The green valley running up the Chisos Mountains just to the right of Crown Mountain is called Pine Canyon. This peaceful scene belies the fact that Pine Canyon is a volcanic caldera, formed after eruptions emptied a large underground magma chamber and the volcano collapsed. The Pine Canyon Caldera measures about four miles in diameter. The beautiful peaks (left to right) of Crown Mountain, Panther Peak, and Pummel Peak are intrusions of high-silica magma called rhyolite, revealed by later erosion.

▼41. Nugent Mountain

Big Bend National Park

Nugent Mountain is a cooled, crystallized magma chamber that was formed when molten rock spread in strata below the surface and then lifted and cracked the earth above. After the surrounding materials were eroded, the solid igneous rock that we see today remains. This area has unusual magnetic properties, revealed during aeromagnetic mapping by geologists. The top of this mountain was the site of an infamous murder.

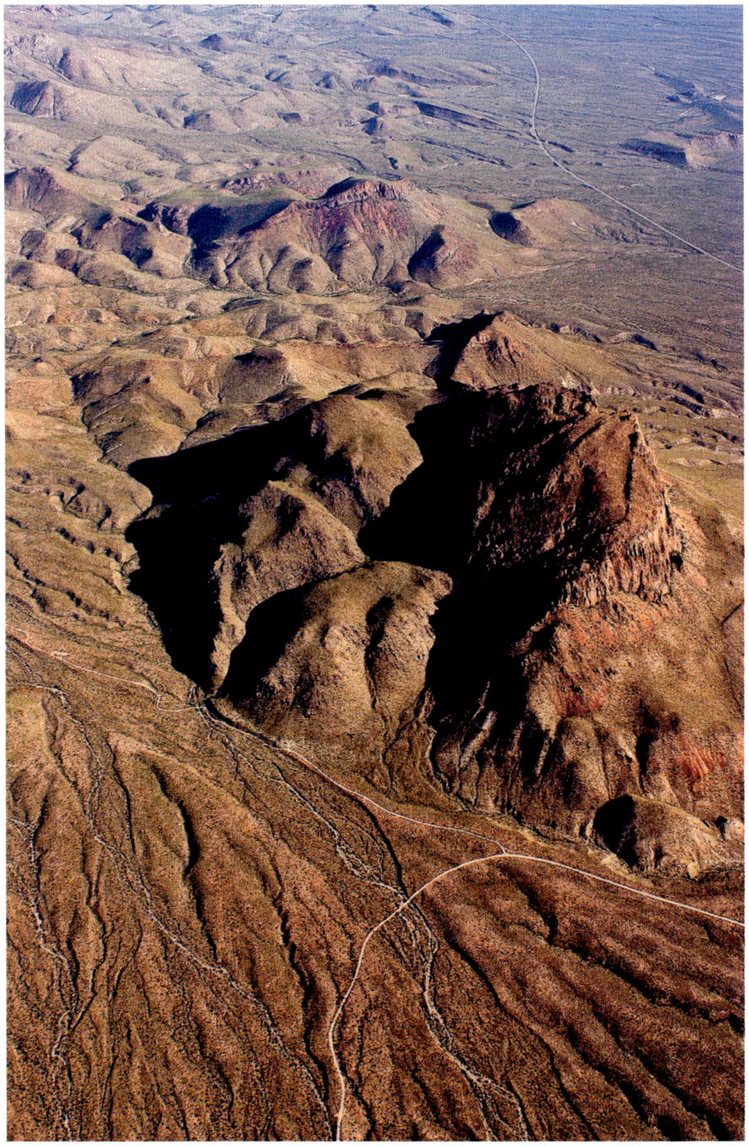

▶42. Through the Window

Chisos Mountains

This is the view looking east through the Window between Vernon Bailey Peak and Ward Mountain into the Chisos Basin, where the Chisos Lodge, restaurant, and campsites are located. At the top (left to right) are Pulliam Ridge, Panther Peak, Pummel Peak, Lost Mine Peak, Casa Grande, and the South Rim. Here the misty haze is backlit by the rising sun, creating ethereal majesty. The mystical mood is fitting for these legendary mountains.

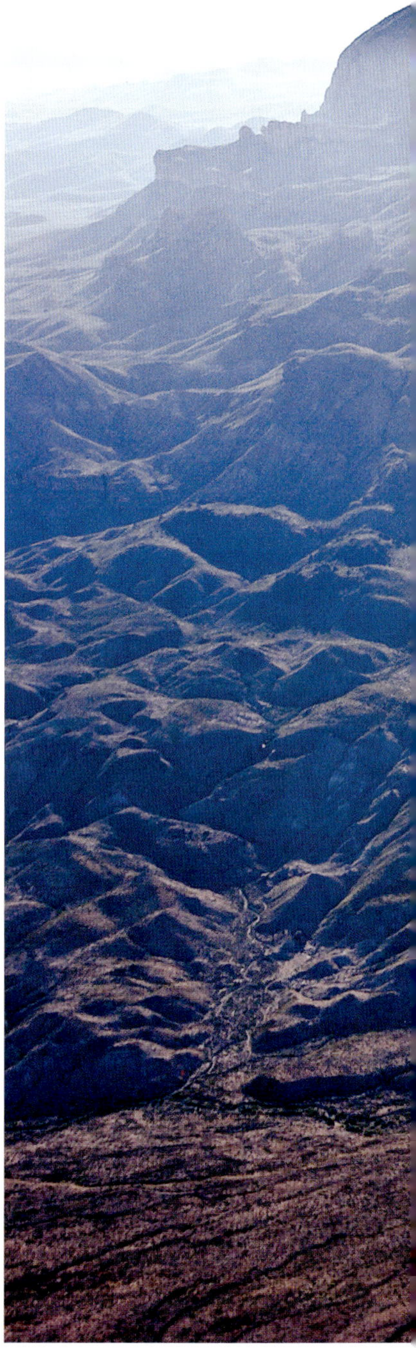

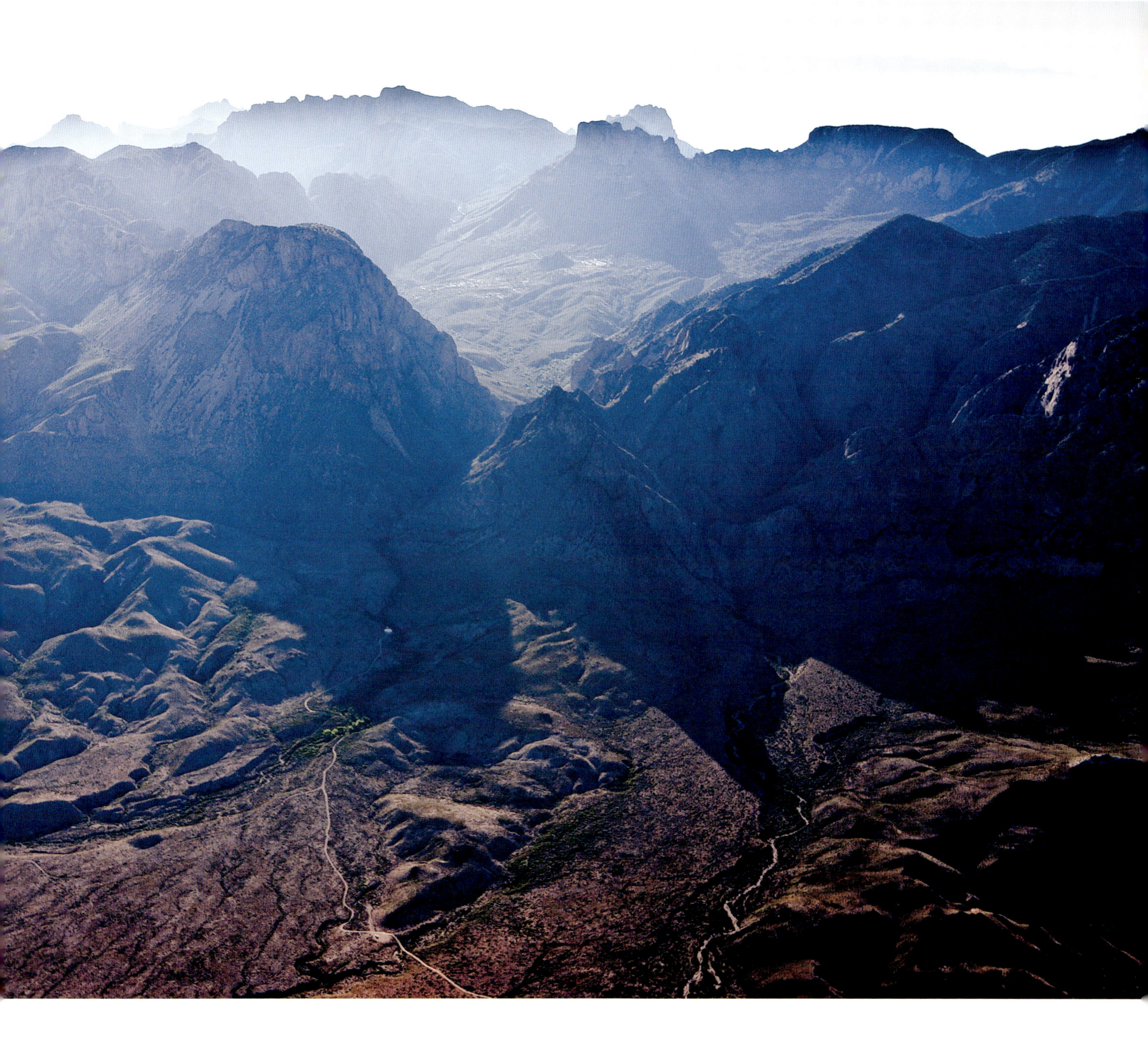

IV. LOOKING FOR WATER

Across Tornillo Flats to Ernst Tinaja

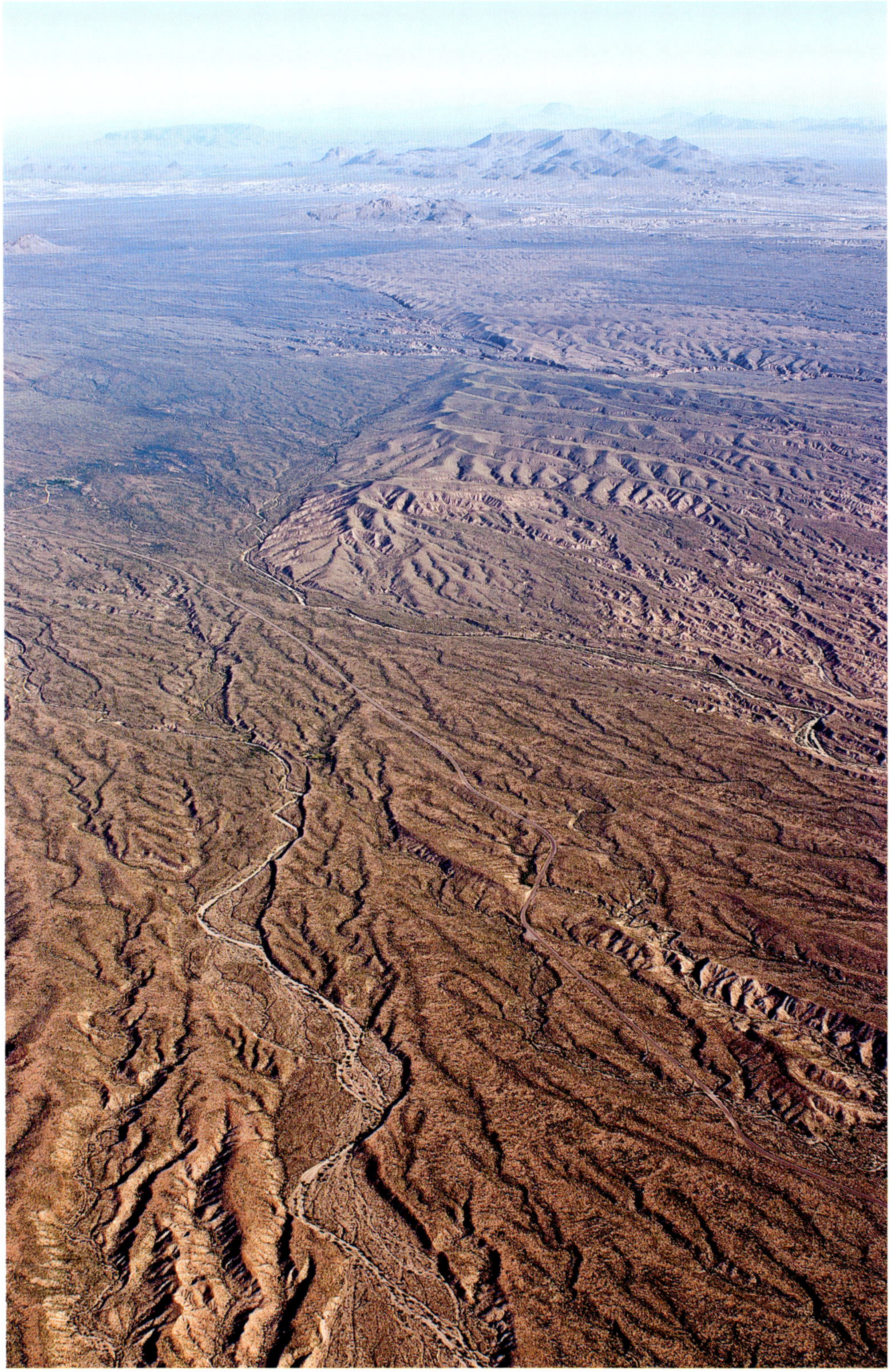

43. Dugout Wells

Estufa Hills

In the desert, water is life. A dot of green in a sea of brown, the small oasis known as Dugout Wells (left) offered water from a spring and the shade of cottonwood trees to weary travelers. Seven miles from Panther Junction, the place remains a sanctuary for desert wildlife and is a prime spot for bird watching. In the early 1900s a small community formed around the well, and a small school was built. The windmill and ruins remain today. From bottom to top are Dugout Draw, Park Route 12, the Estufa Hills, and the Rosillos Mountains.

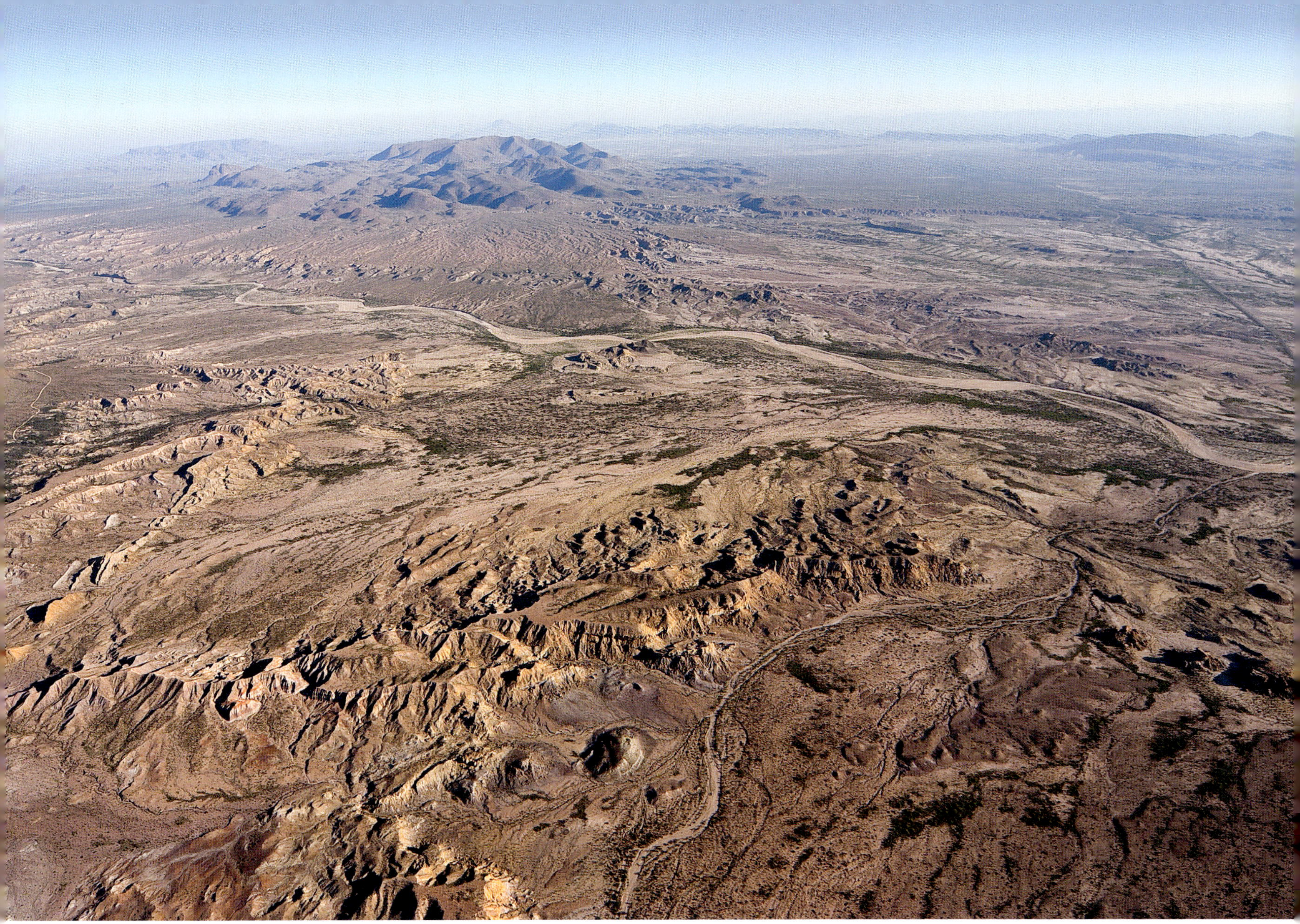

44. Tornillo Creek

Chihuahuan Desert southeast of Panther Junction

Beneath the Rosillos Mountains sometimes flows Tornillo Creek, named after the screwbean mesquite tree, or *tornillo* in Spanish. The twisting creek bed, often dry, turns into a brown torrent in flash floods following desert thunderstorms even miles away. The surrounding Tornillo Flat was the main route of the Great Comanche War Trail, which passed through Persimmon Gap, visible in the far distance to the north. Today Texas Highway 385 follows a similar path through the natural pass in the Santiago Mountains. The openness of this winsome desert scene lured me, and I included this picture to communicate the immensity of this expansive no-man's-land.

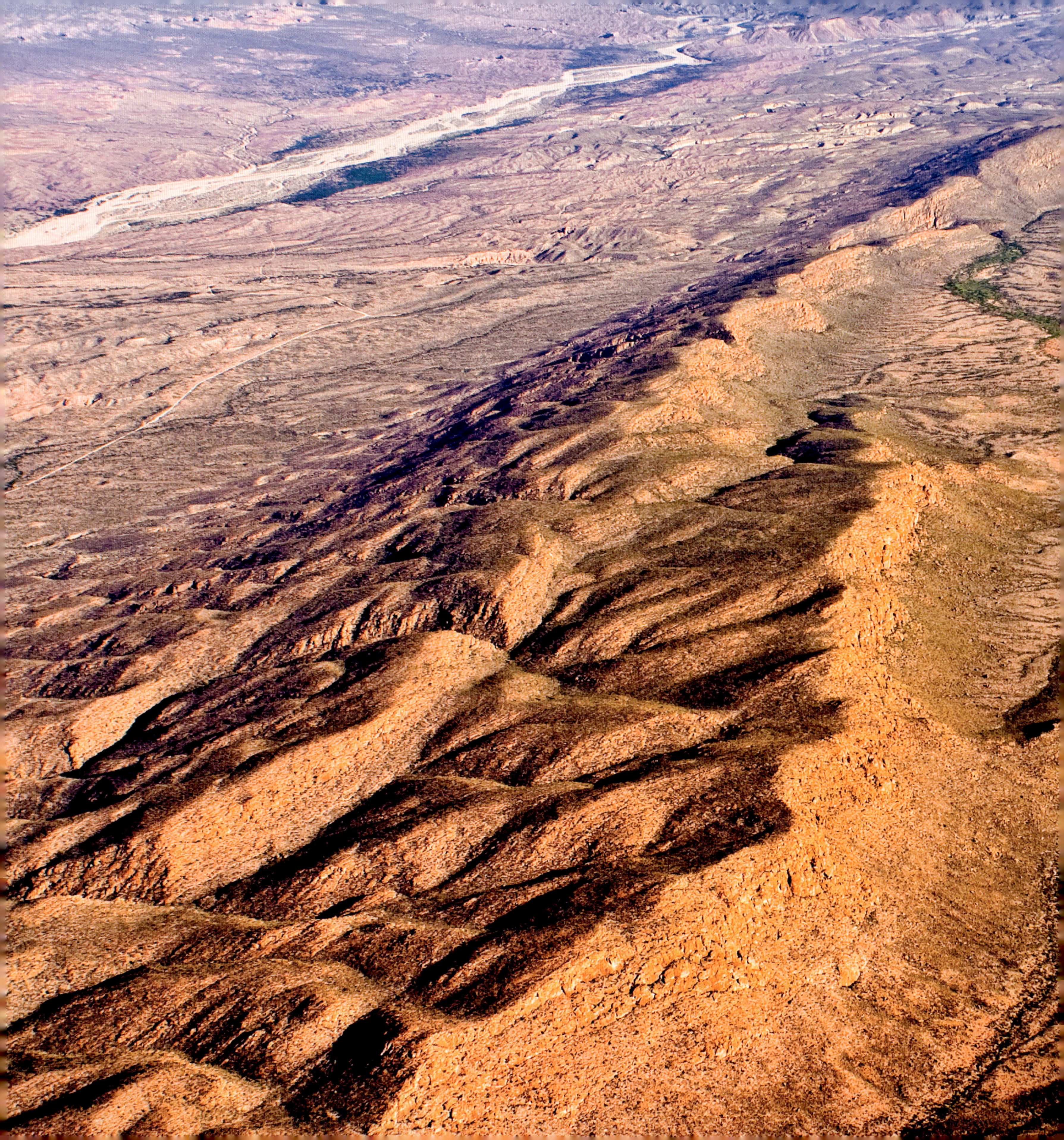

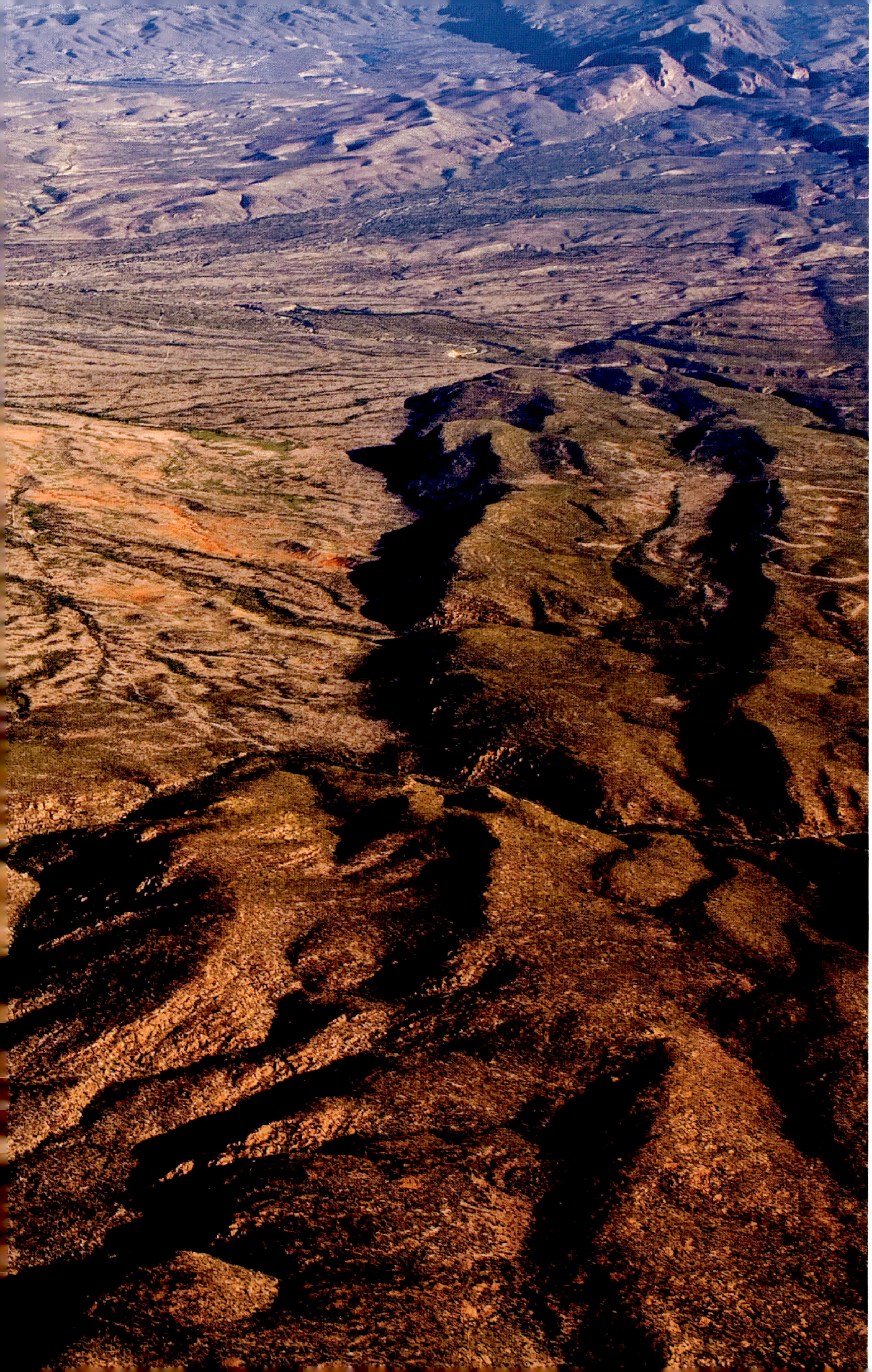

45. The Land of Ernst

Seven miles northwest of Boquillas

From right to left are Ernst Basin, Ernst Tinaja, Tornillo Flat, and Tornillo Creek. Many landmarks in the area were named after Max A. Ernst, a German immigrant who leased land in this area that included a water hole named the Big Tinaja. Ernst lived in La Noria, which was located along Tornillo Creek (left). Ernst wore many hats, including notary public, justice of the peace, coroner, rancher, shop owner, county commissioner, and postmaster of Boquillas. He was murdered by gunshot in 1908.

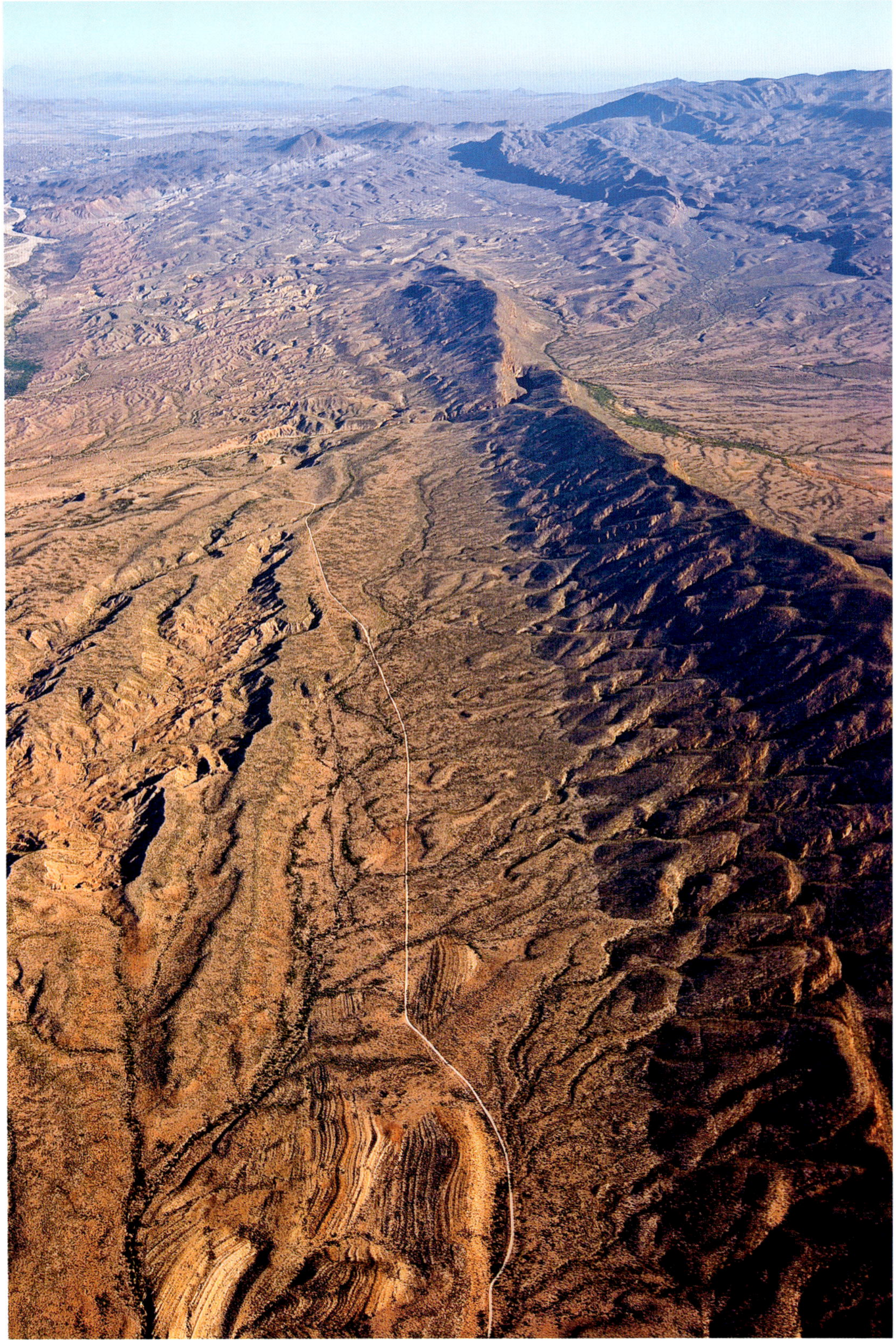

46. Old Ore Road

West of Ernst Tinaja

The Old Ore Road is seen as a thread of white running from top to bottom in this view. The road follows the path used in the early 1900s by pack mule trains to transport silver and lead ore from mines in the Sierra del Carmen, Mexico, to the Marathon, Texas, railroad station. Although Ernst Tinaja is hardly visited by tourists, it is one of my favorite places due to the natural beauty as seen from the air, especially the alligator-like ridge named Cuesta Carlota.

47. Ernst Tinaja

Six miles north of Rio Grande Village

Tinaja, the Spanish word for jar, is the term used for natural depressions in the bedrock of stream beds that hold water. In this way, even when a desert stream has no flow, the tinaja stores life-giving water and often supports a unique ecosystem. For ages both people and wildlife have used these geologic cisterns to survive their treks through the desert. The tinajas can also turn into traps for unwary animals that fall into the sunken pools and cannot escape.

Ernst Tinaja is the largest of a series of tinajas in this area. It was formed by erosion in an intermittently flowing creek bed that passes through Cuesta Carlota. *Cuesta,* Spanish for slope, is the term for a ridge with a steep side and a gentle side. Cuesta Carlota, which runs north to south, was formed by a fault line passing along the east side of the ridge, thereby making a steeper slope on one side.

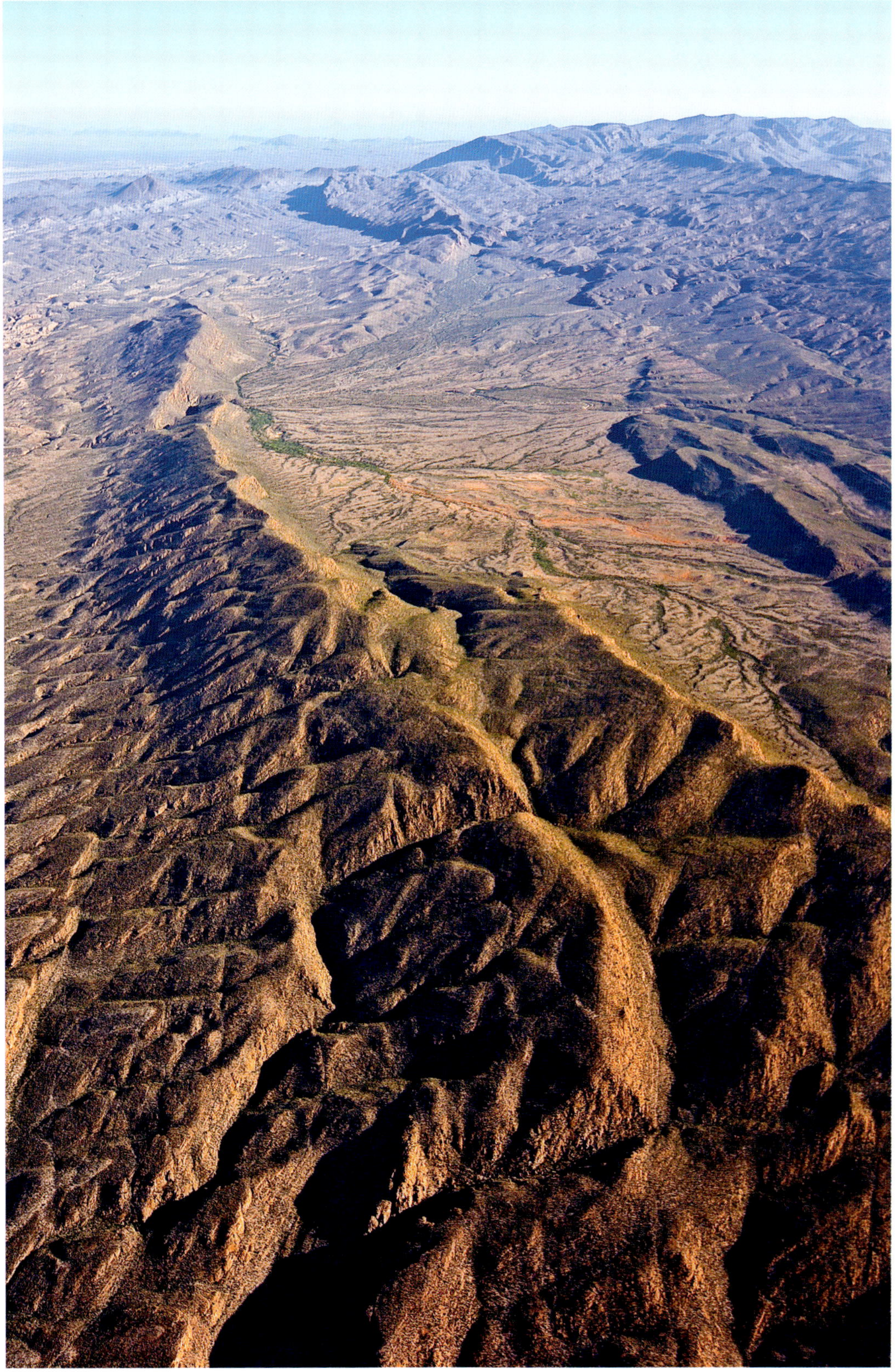

48. Ernst Valley

Four miles north of Boquillas

To the right is Ernst Valley and to the left is Ernst Basin. Ruins of the Ore Terminal Aerial Tramway are located in the lower left corner. The tramway buckets carried silver and lead ore from mines in the Sierra del Carmen across the Rio Grande. From the end of the tramway, teamsters freighted the ore in wagons to the railroad at Marathon, Texas. The ingenious aerial tramway utilized a moving steel cable supported by timber towers to transport ore across the river and the rugged countryside.

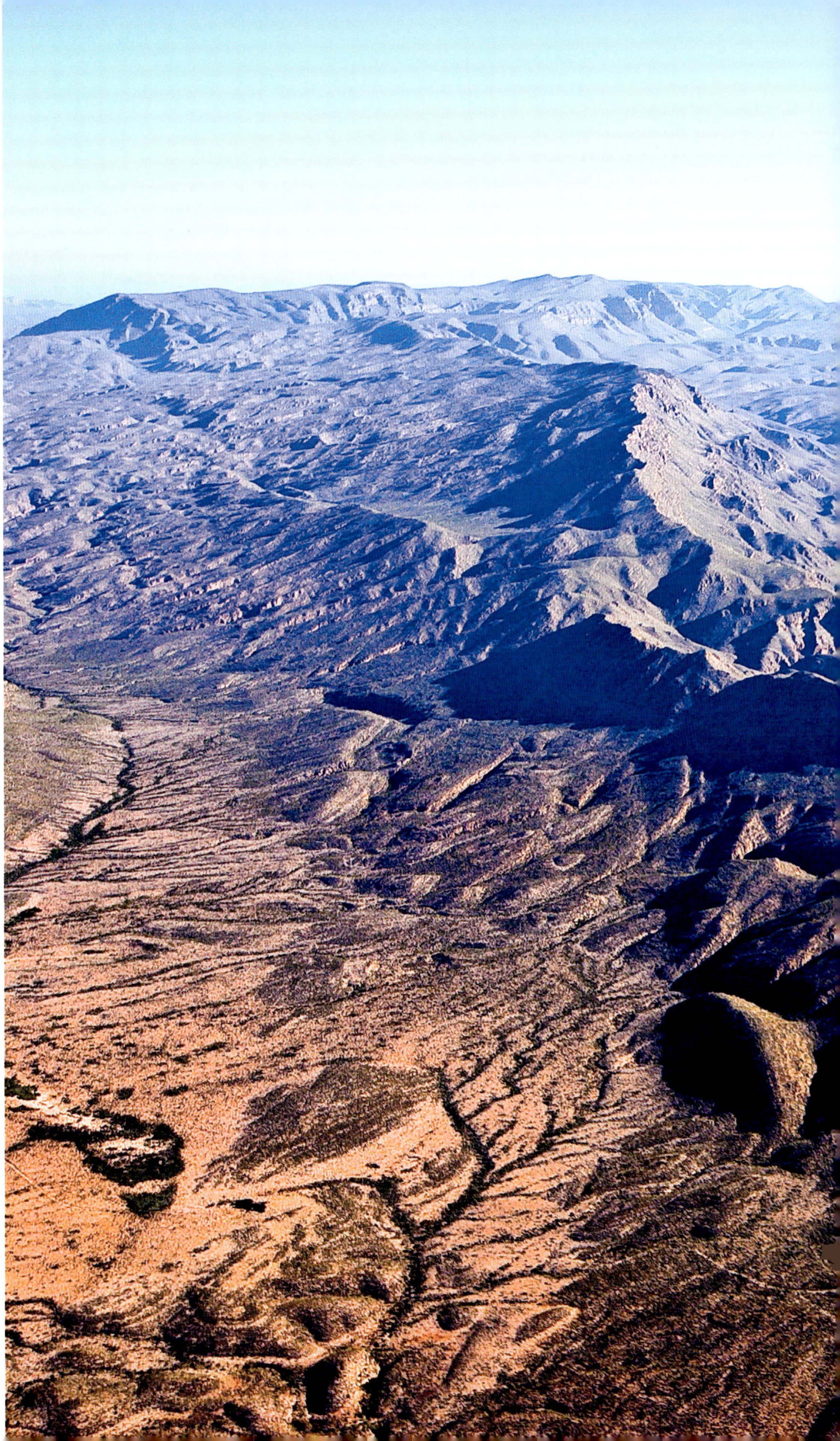

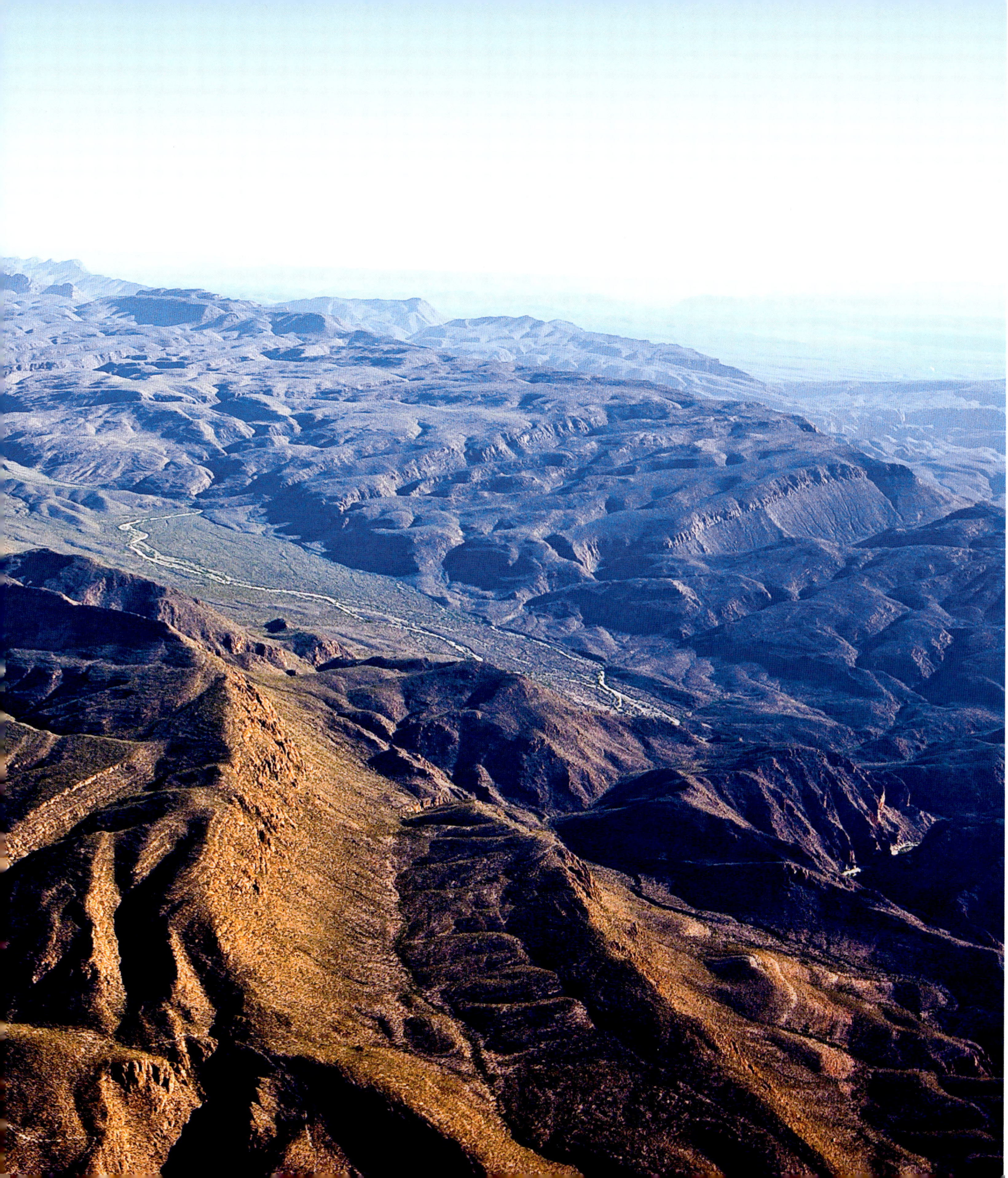

V. THE RIO GRANDE

Flying Upstream to Smoky Creek

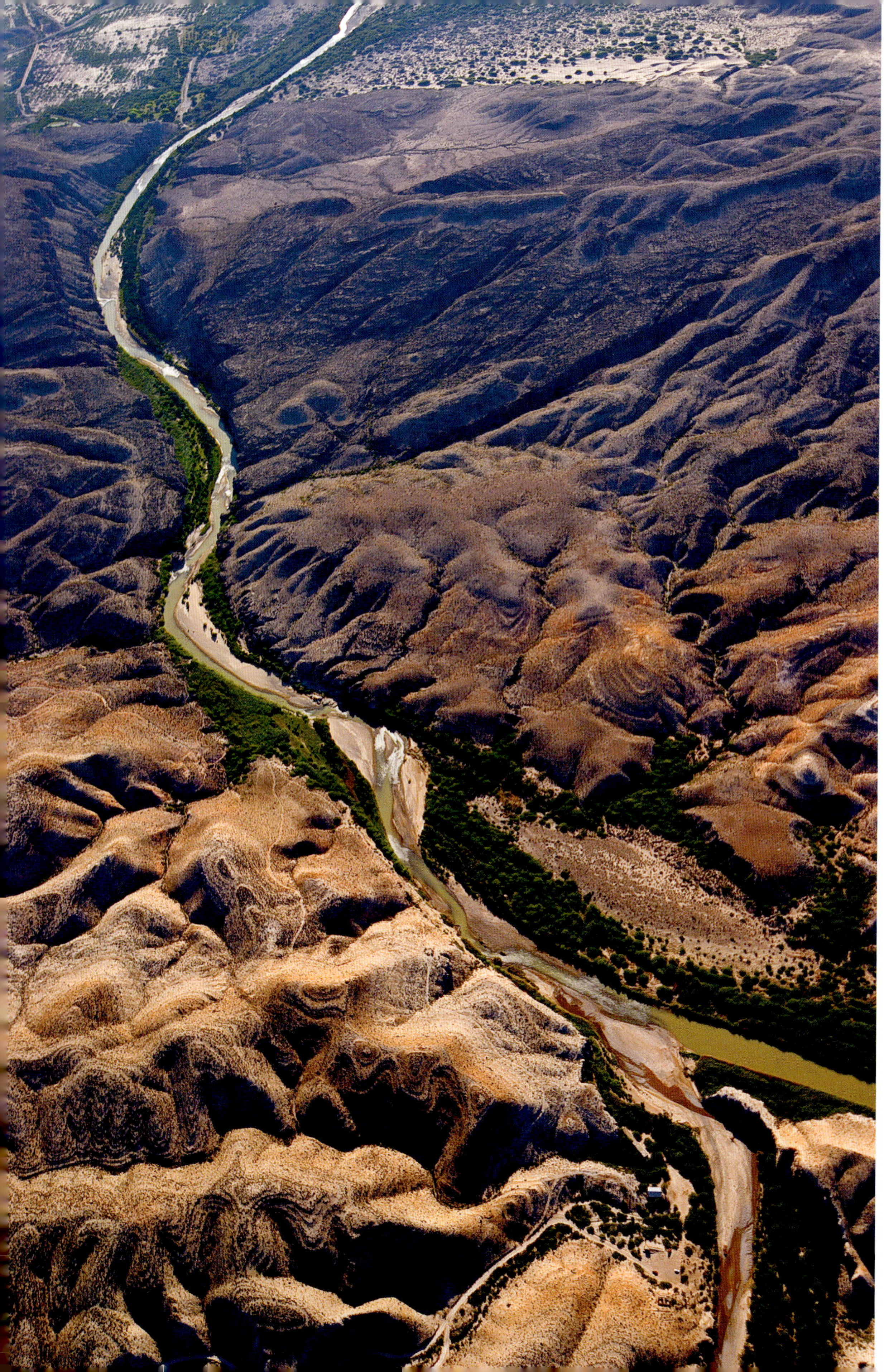

49. Rio Grande, Silver Shining River

Looking toward Rio Grande Village

We had just done a very tight turn with the aircraft to take a photograph looking down on Langford Hot Springs, when I looked to the east and saw the Rio Grande shining silvery in the sunlight. In the distance were Rio Grande Village and the remains of the old cotton fields there. The river formed a winding ying-yang shape as it flowed through the mountains and hills, separating the United States from Mexico. The Rio Grande is a thread of life through the desert, creating a green sash through an arid wasteland. This segment of the river, known as Hot Springs Canyon, is part of the sixty-nine-mile stretch that is a designated National Wild and Scenic River. Sadly, bacteria counts in the river are high due to upstream sewage discharge from populated areas.

50. Langford Hot Springs

Sixteen miles southwest of Panther Junction

Near the confluence of Tornillo Creek and the Rio Grande, the therapeutically soothing waters of Langford Hot Springs are 105 degrees F year round. The groundwater is geothermally heated as it travels deep through faults in the earth's crust, then emerges here. The Boquillas Hot Springs system has a flow rate of about 250,000 gallons per day. This spot is named for J. O. Langford, who came to the springs to regain his health, then claimed the land under the Homestead Act. The remains of his store and the foundation of his once-impressive bathhouse are preserved in the Hot Springs Historic District. The water is irreplaceable ancient fossil water that is rich in minerals and said to have healing powers. Langford Hot Springs are open to the public, and many hikers still stop there to soothe their aching muscles, much like the travelers of long ago. The bold abstract shapes of the land cut by the streams are bewitching to see from above.

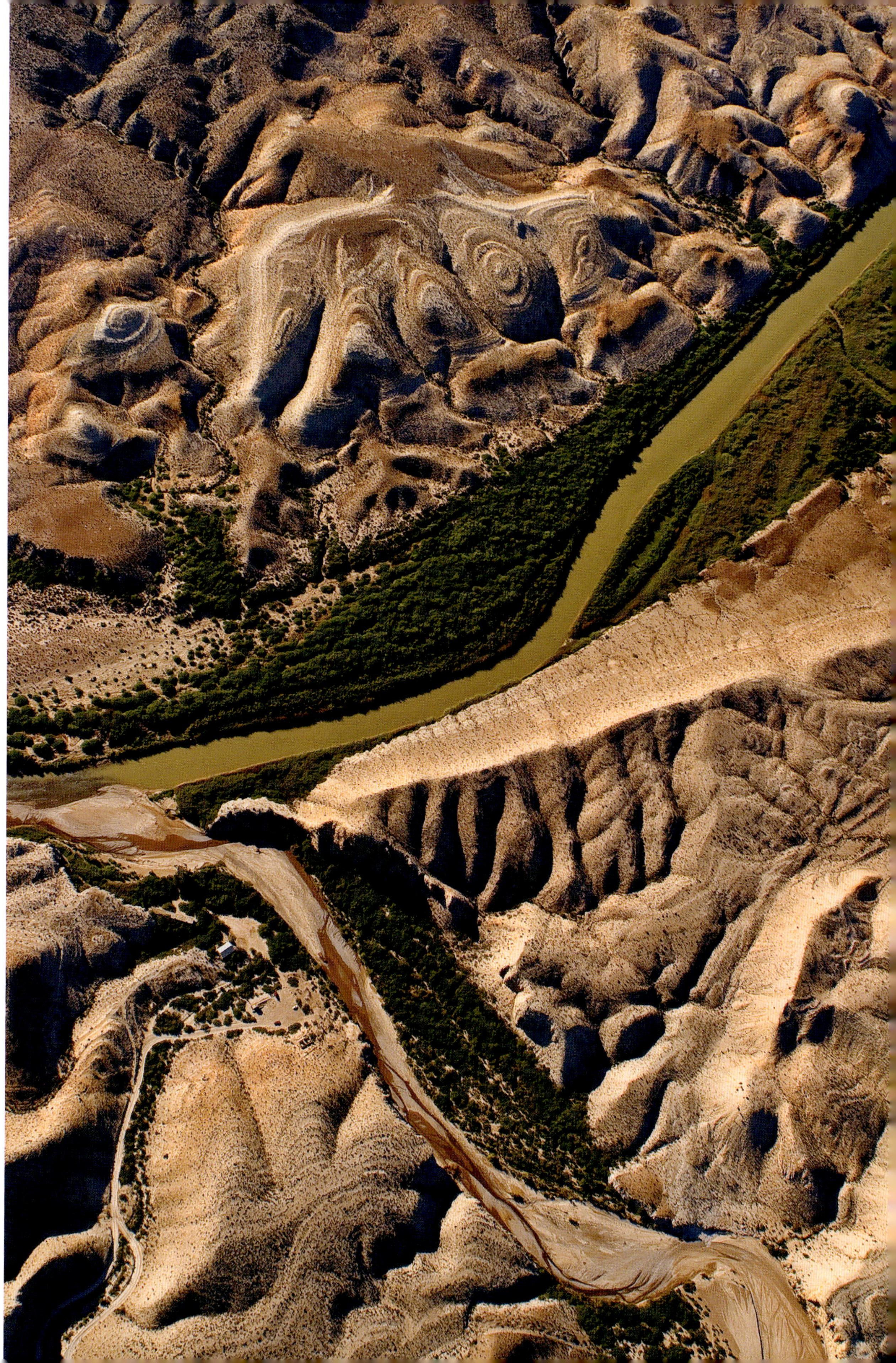

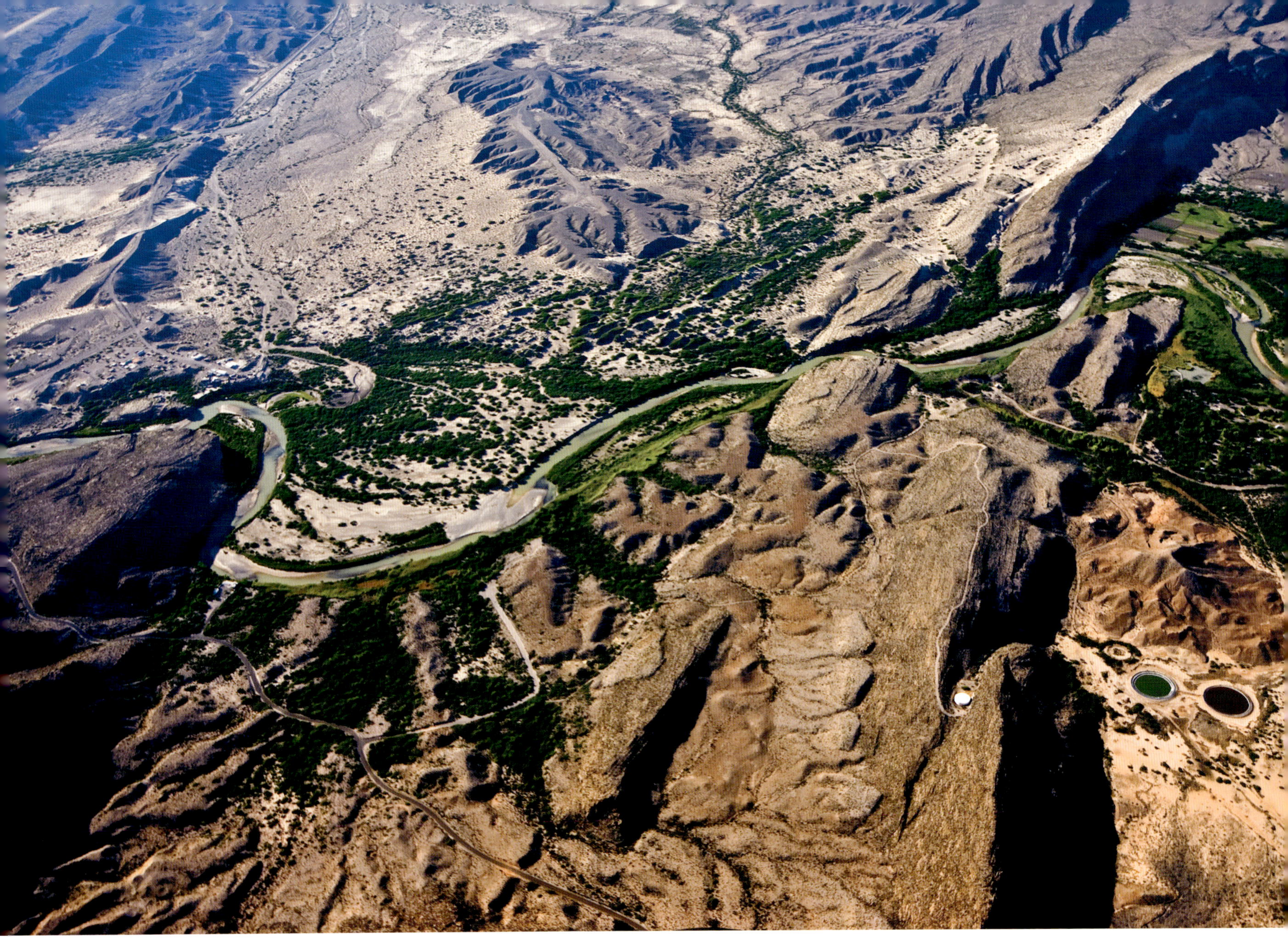

51. Boquillas Crossing

Just east of Rio Grande Village

Many of the traditional river crossings where people freely traveled back and forth across the Rio Grande were closed for security reasons after the events of September 11, 2001. The old silver mining town of Boquillas, Mexico, seen on the other side of the river (far left), depended heavily on tourist dollars and suffered greatly from the crossing closure. In April 2013 US Customs and Border Protection and the National Park Service reopened the Boquillas port of entry. Visitors can cross the river in small boats, including one operated by a legendary singing Mexican boatman who is now back in business.

52. Rio Grande Village

Big Bend National Park

Rio Grande Village was once the cultivated agricultural land of Daniels' Ranch, irrigated with water from the Rio Grande. Just a stone's throw from Mexico, the area is now the site of a national park campground. A nature trail allows visitors to traverse the lush wetlands along the river that teem with life, including beavers that have created beaver ponds there. The green area along the river, known as the riparian zone, is one of the best year-round birding locations in the national park. The circular tanks (center) of green and brown water below the ridge are part of a water treatment facility. I composed this image to capture the graceful crescent-shaped ridge, known as the Sierra el Oro on the Mexican side, that is so visually dominant when seen from above—but hard to perceive at ground level.

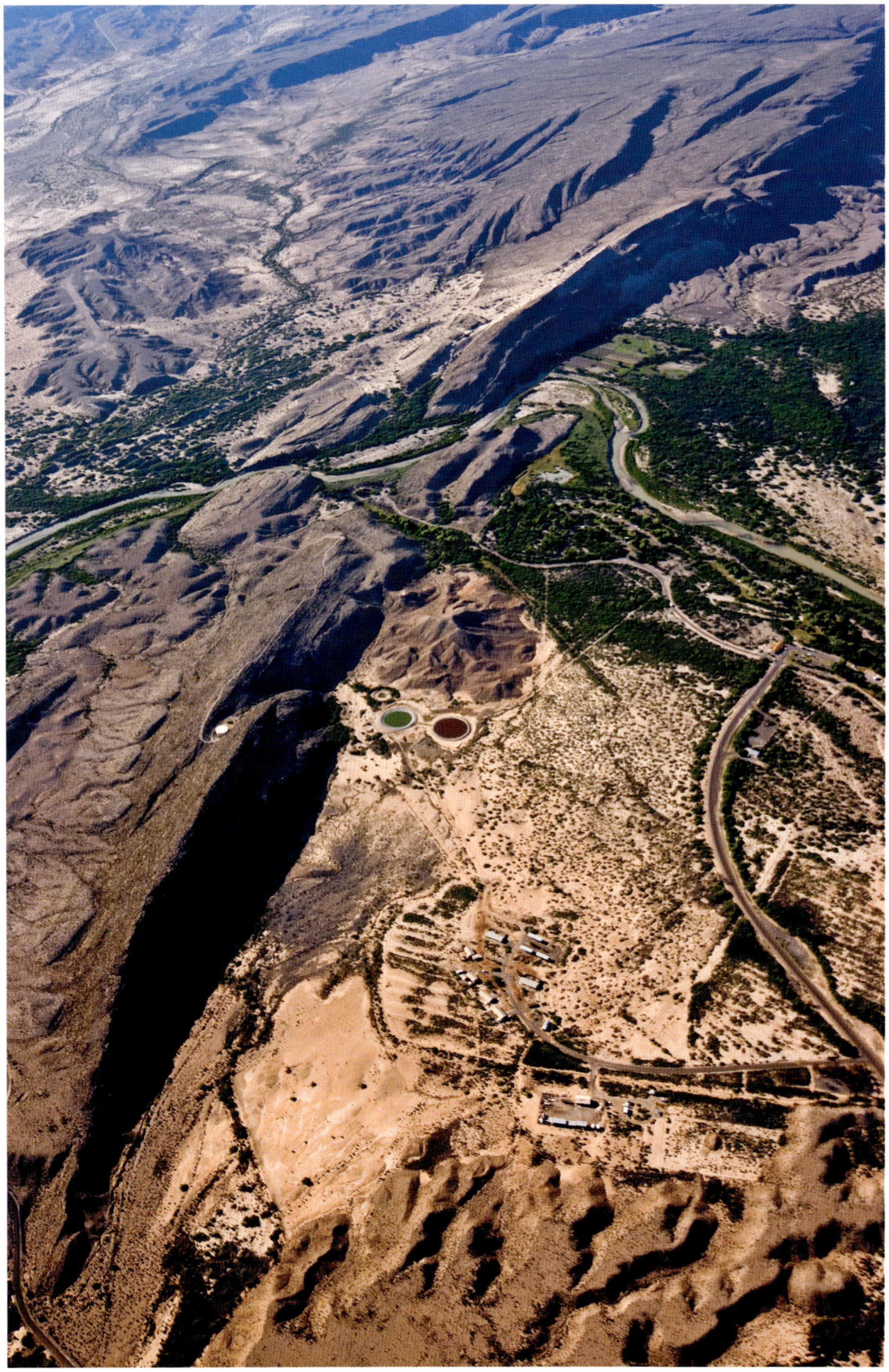

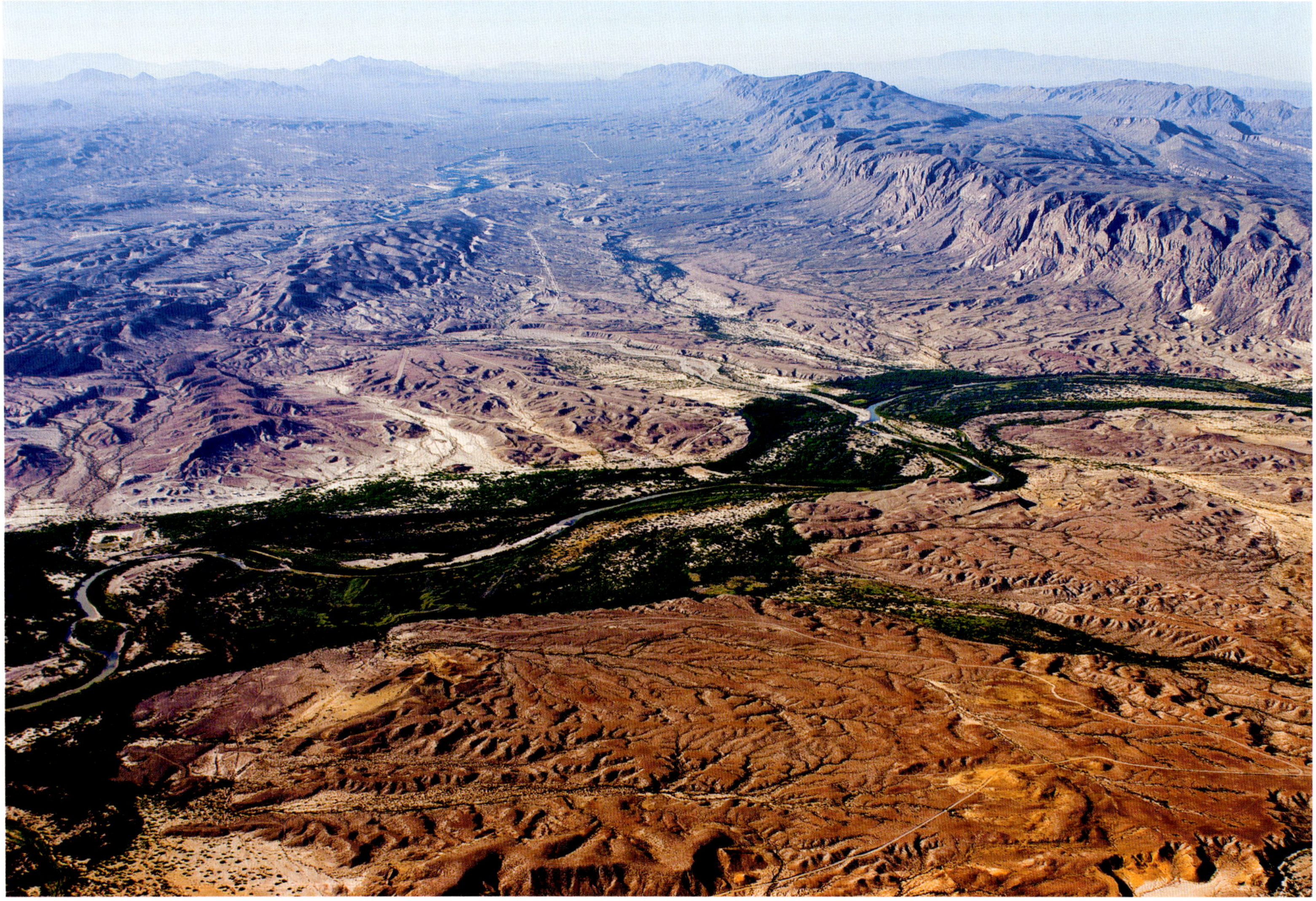

53. San Vicente Panorama

Along the Rio Grande

Present-day San Vicente, Coahuila, Mexico (center), is located across the Rio Grande near the site of the Presidio de San Vicente, a Spanish fort built in 1774 to protect the area from Comanche, Apache, and bandit raids. San Vicente Crossing (left) was used throughout history as a point to ford the river until it was closed in 2002 as part of the US effort to increase border security. I favor this photo for how it presents this area of rich borderland history using a wide-angle lens to reveal the stirring gamut of geography from the green belt along the river to the thrusting uplift of the Sierra San Vicente (right).

54. The Road to San Vicente Crossing

North of San Vicente, Mexico

The verdant riverbanks of the Rio Grande (top) form a striking contrast to the stark earth colors of the desert arroyo (center). In the middle ground is River Road East, which winds along the Rio Grande on the US side of the border. Smugglers cross these desert badlands carrying seventy-five-pound packs of contraband including drugs, weapons, and explosives. I elected to include this more detailed view of the erosion, taken at low altitude, for its astonishing figures that water has carved out of the desert.

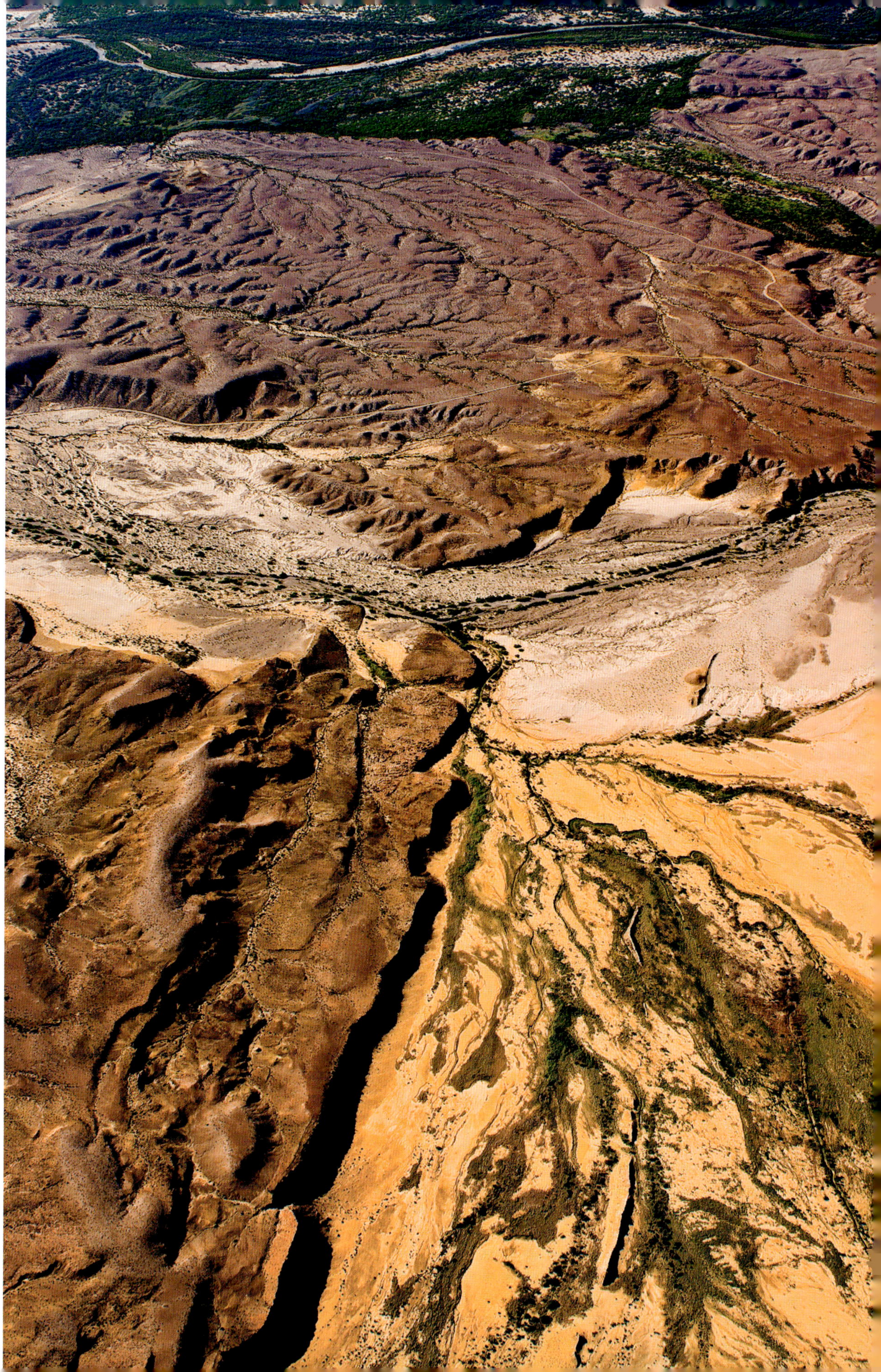

55. Gazing toward San Vicente, Mexico

Upstream from Rio Grande Village

Looking southeast toward Mexico, one sees the Rio Grande emerge from San Vicente Canyon (right). In the foreground the eroded alluvial plain of the Chihuahuan Desert forms an otherworldly landscape of badlands. At the base of the Sierra San Vicente, on the US side of the Rio Grande, is the Compton place, named for C. G. Compton, a local storekeeper and farmer whose son was killed by bandits in 1916. The turn in the river here is a longtime favorite fishing spot.

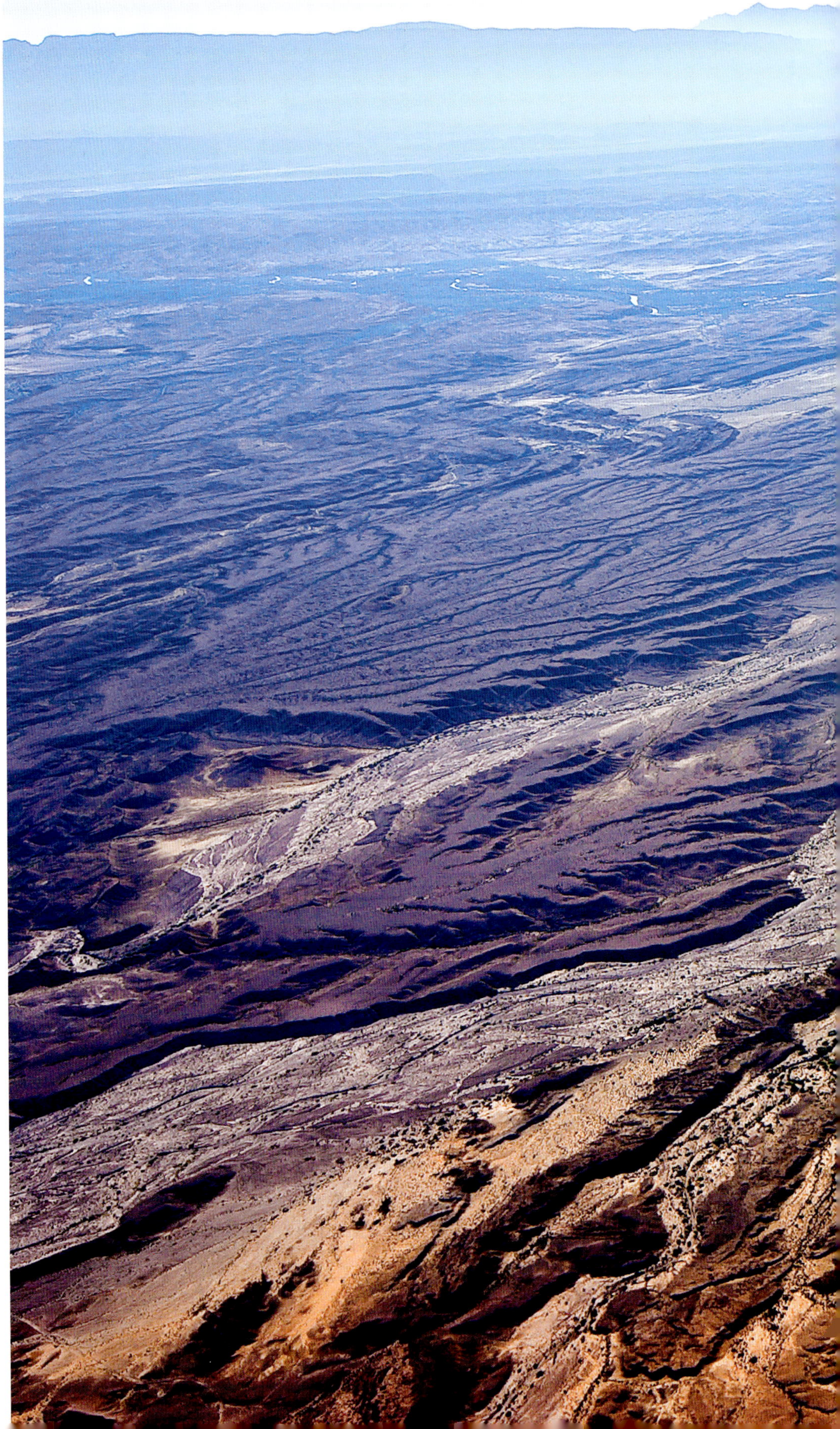

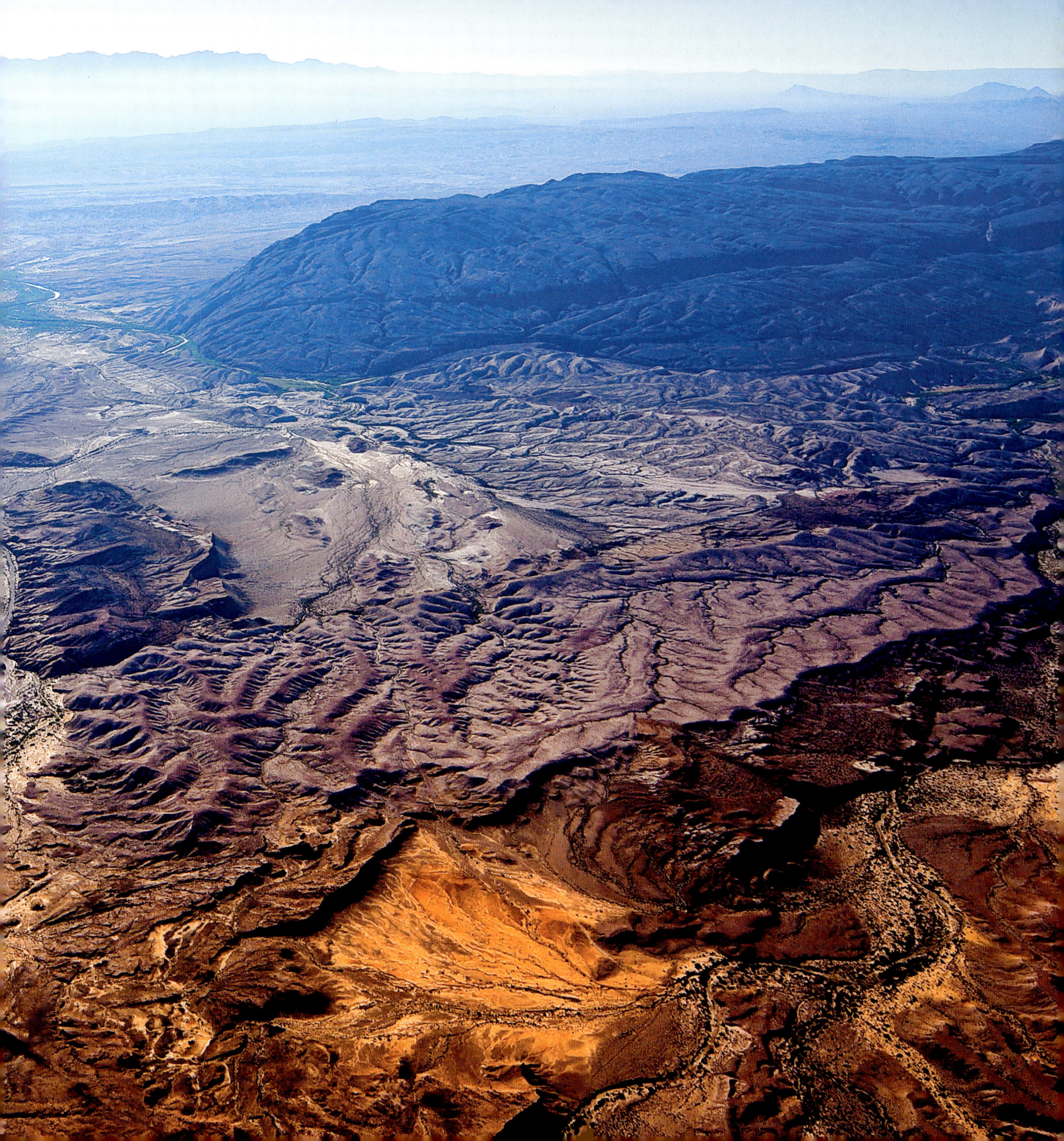

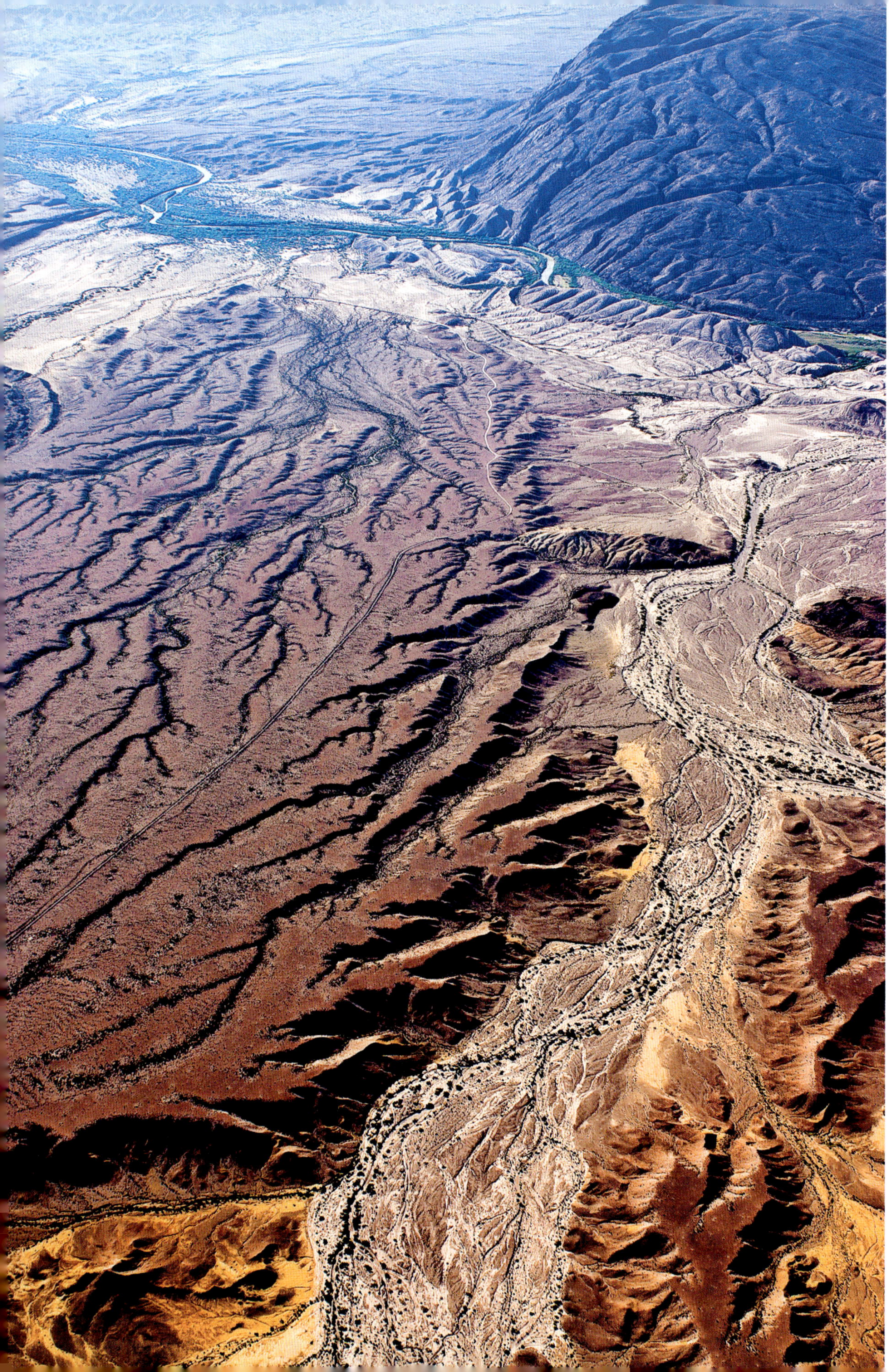

56. Rooney's Place

Glenn Draw

The Rio Grande flows in front of the monolithic dark mass of the Sierra San Vicente anticline. Below the tip of the massive range lies Rooney's Place (center), a tumbled ruin of the home of Francis Rooney, a Big Bend rancher. In the foreground are the deep ravines formed by erosion around Glenn Draw. To the left is the road to the Casa Piedra site, now a national park campsite. Dramatic views such as this one, which boldly show the marks on the land made by natural forces, often appealed to me more than well-known landmarks. This image remains almost surreal to me with the staggering scale and textures of the land; it is one of my favorites.

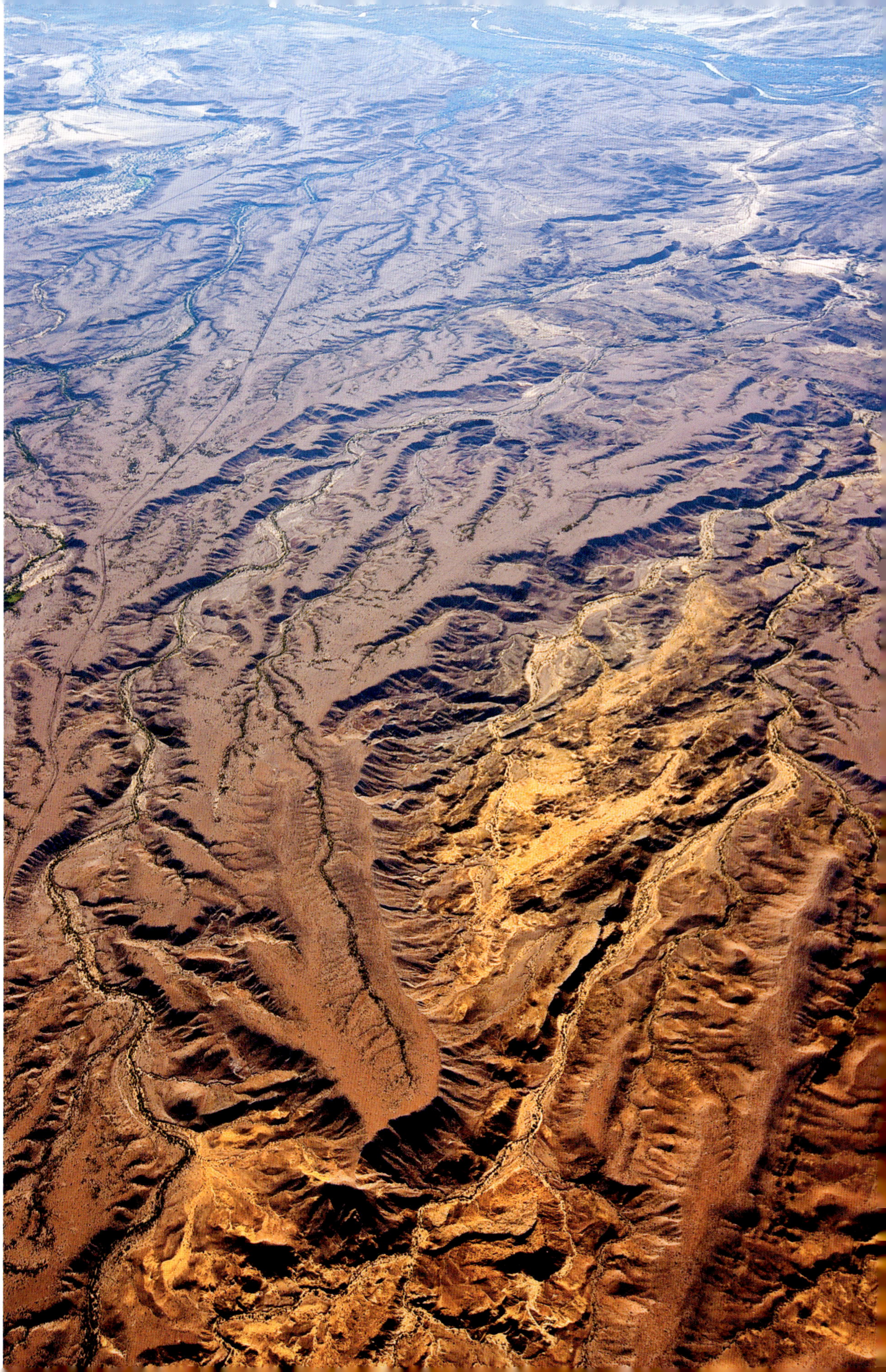

57. Badlands

Near San Vicente

Here the arroyos stretch toward hilltops like dry veins in a brown leaf hoping for rain. These gullied alluvial plains stand witness to the effects of erosion in areas where the topsoil has been washed away by flash floods and in some cases triggered by overgrazing. This patch is located between River Road East and Glenn Springs Road, about three miles north of G. C. Compton's former store. In the distance is the Rio Grande (top), painting a border of green through the desert.

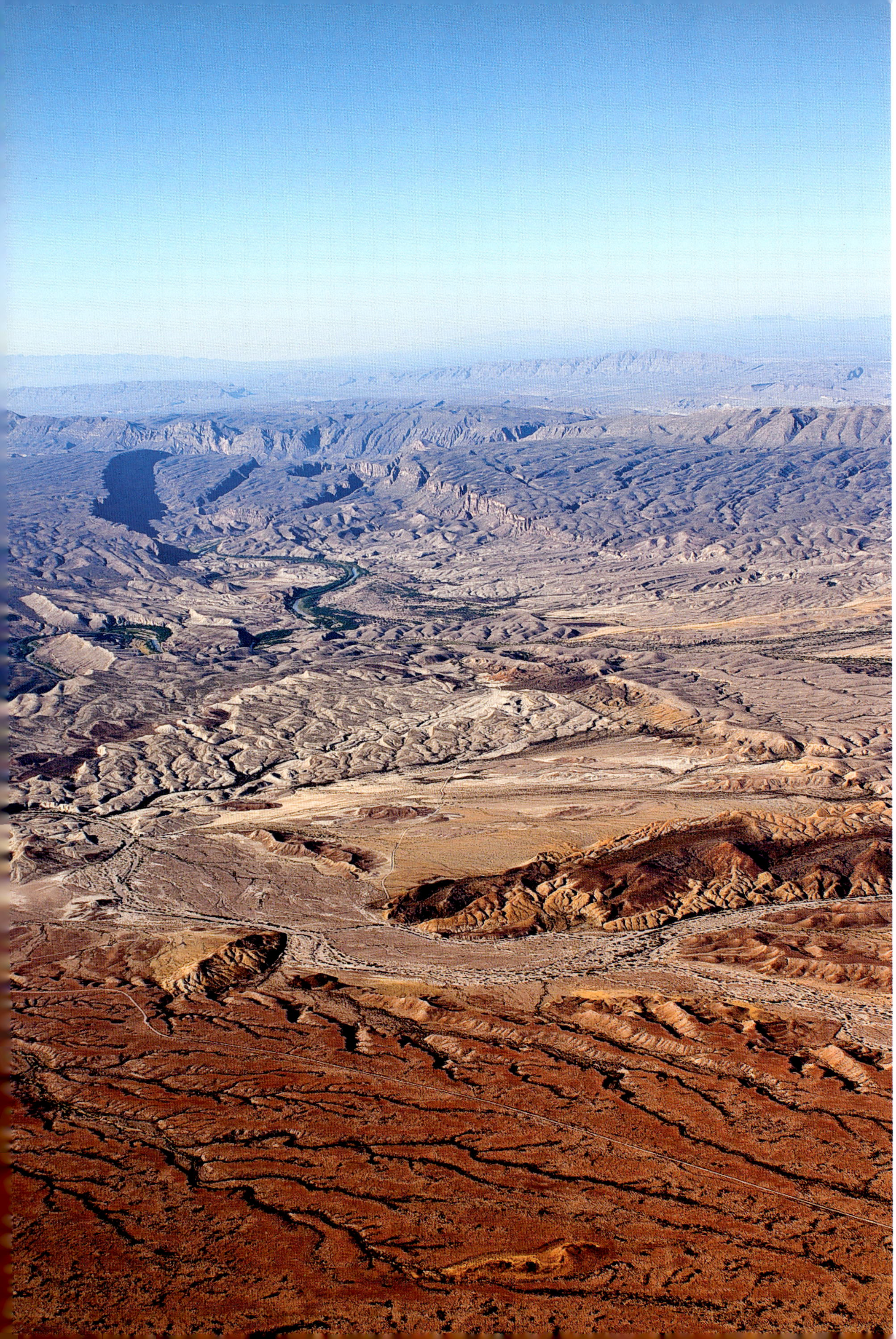

58. Solis

Downstream mouth of Mariscal Canyon

A sash of green and blue marks the Rio Grande at Solis Landing (center), where it exits Mariscal Canyon. Recreational rafters and kayakers who float the river downstream through the canyon frequent the area. Besides White Rock rapid at the canyon mouth, there are other dangers. This area is locally known as a crossing for drug smugglers known as "mules." There have been several brutal murders along the Rio Grande, including here, and the National Park Service advises that visitors not camp where Mexican roads approach the river.

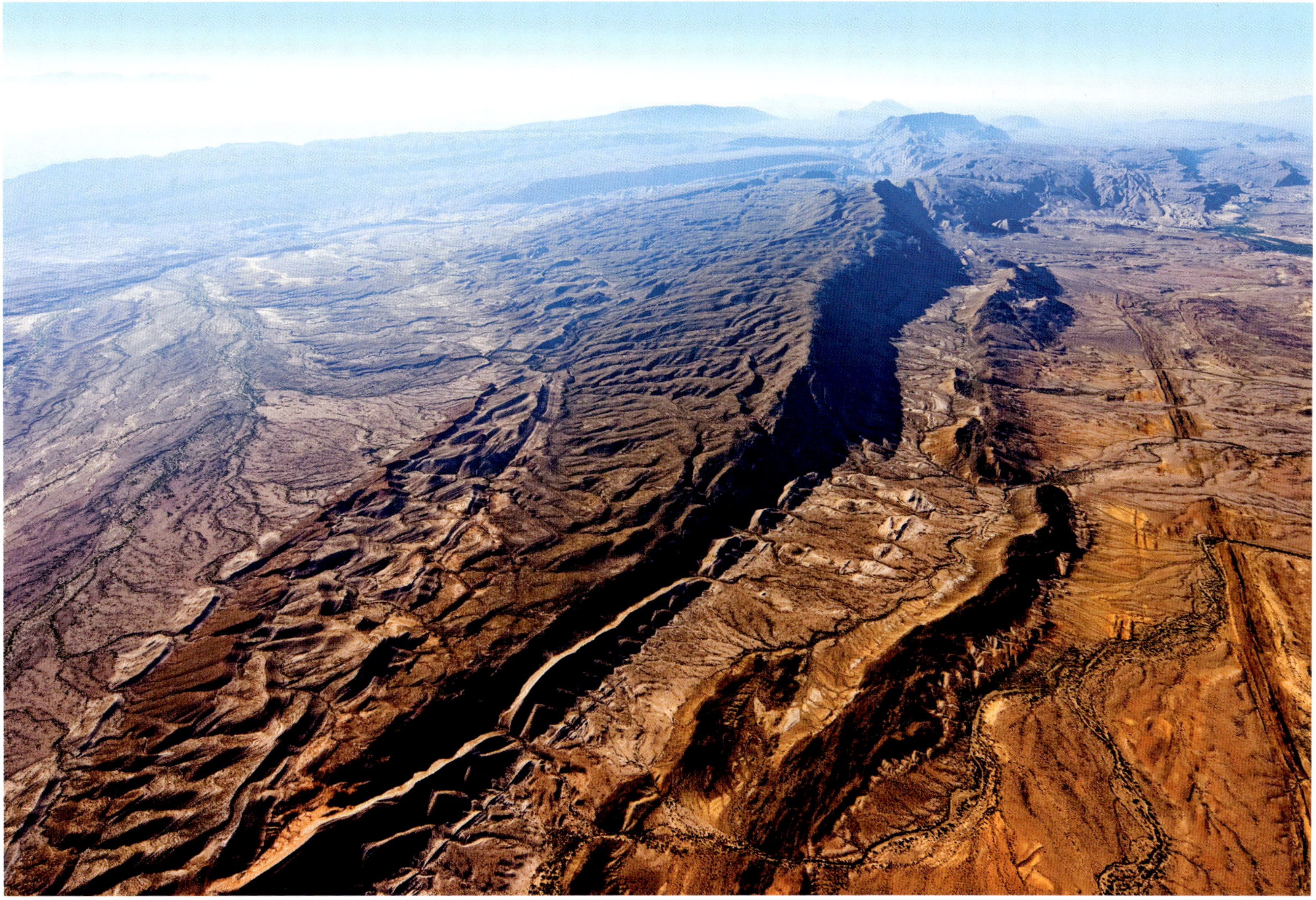

59. Mariscal Mountain

Seventeen miles southwest of Boquillas

Mariscal Mountain measures about one mile wide and nine miles long. It is an anticline or upward folding of the earth's crust. Mariscal Mountain was created by the same forces as the Rocky Mountains and is considered to be the southernmost US extension of that range. The wide uplift of Cretaceous limestone created the rugged cliffs seen today. Mariscal Canyon of the Rio Grande is visible in the distance (top center). The Mariscal cinnabar mine once operated on the north end. The ore was extracted from mine shafts in the mountain (lower left), then processed using furnaces and condensers. During the first half of the twentieth century, the mine produced about 1,400 flasks of mercury that weighed seventy-six pounds each. The mercury was used heavily during World War I. During its heyday, the Mariscal Mine produced almost 25 percent of all the mercury in the United States. Miners and smelter operators were paid between $1.00 and $1.50 per day and undoubtedly suffered from the effects of mercury poisoning. The area is still contaminated. To the left is Fresno Draw and Solis Landing Road, both of which run south to the Rio Grande. In terms of visual interest, I selected this photograph for the rhythmic lines of the ridges, which seem to have a free-flowing cadence as they move away from the camera.

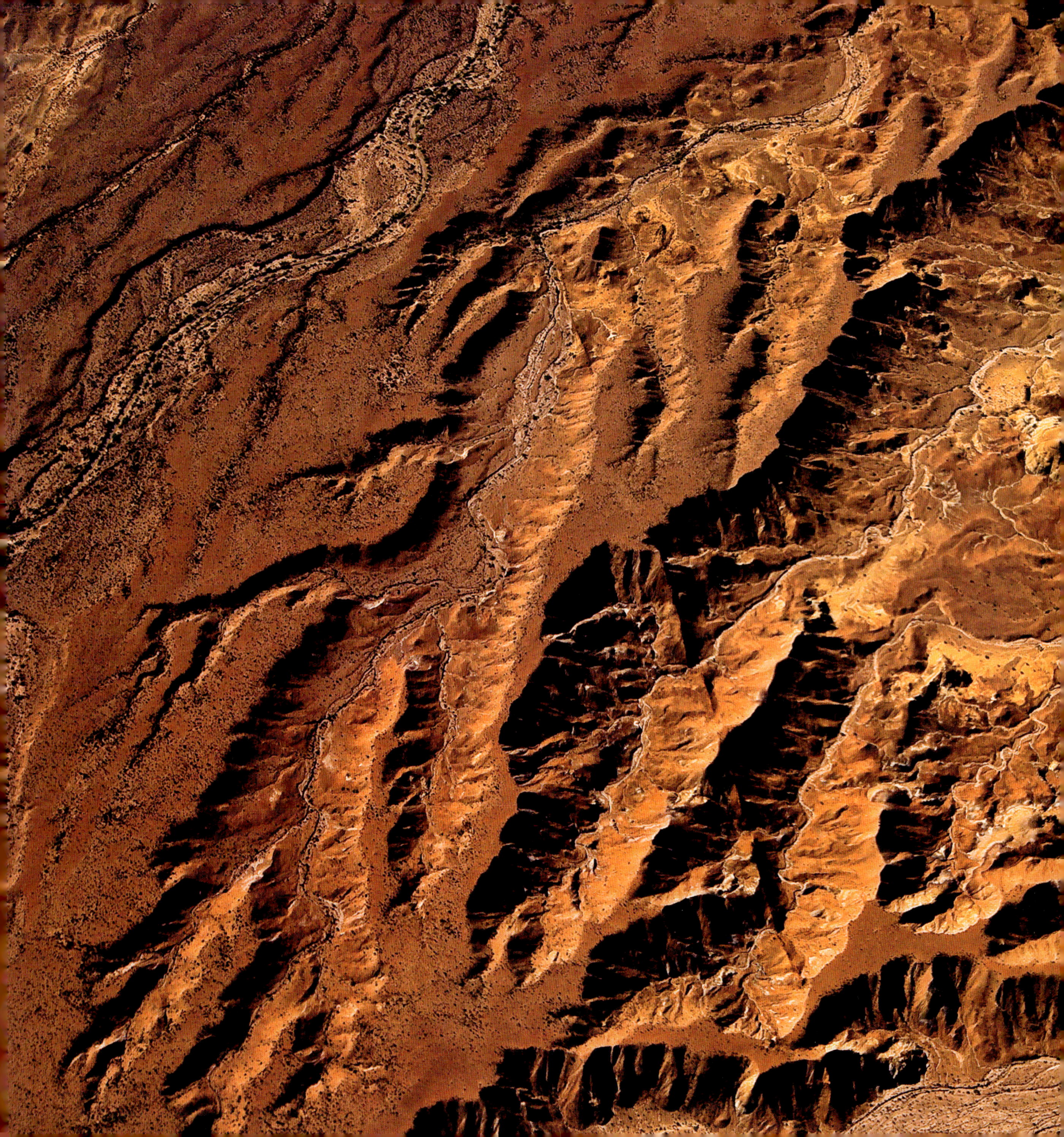

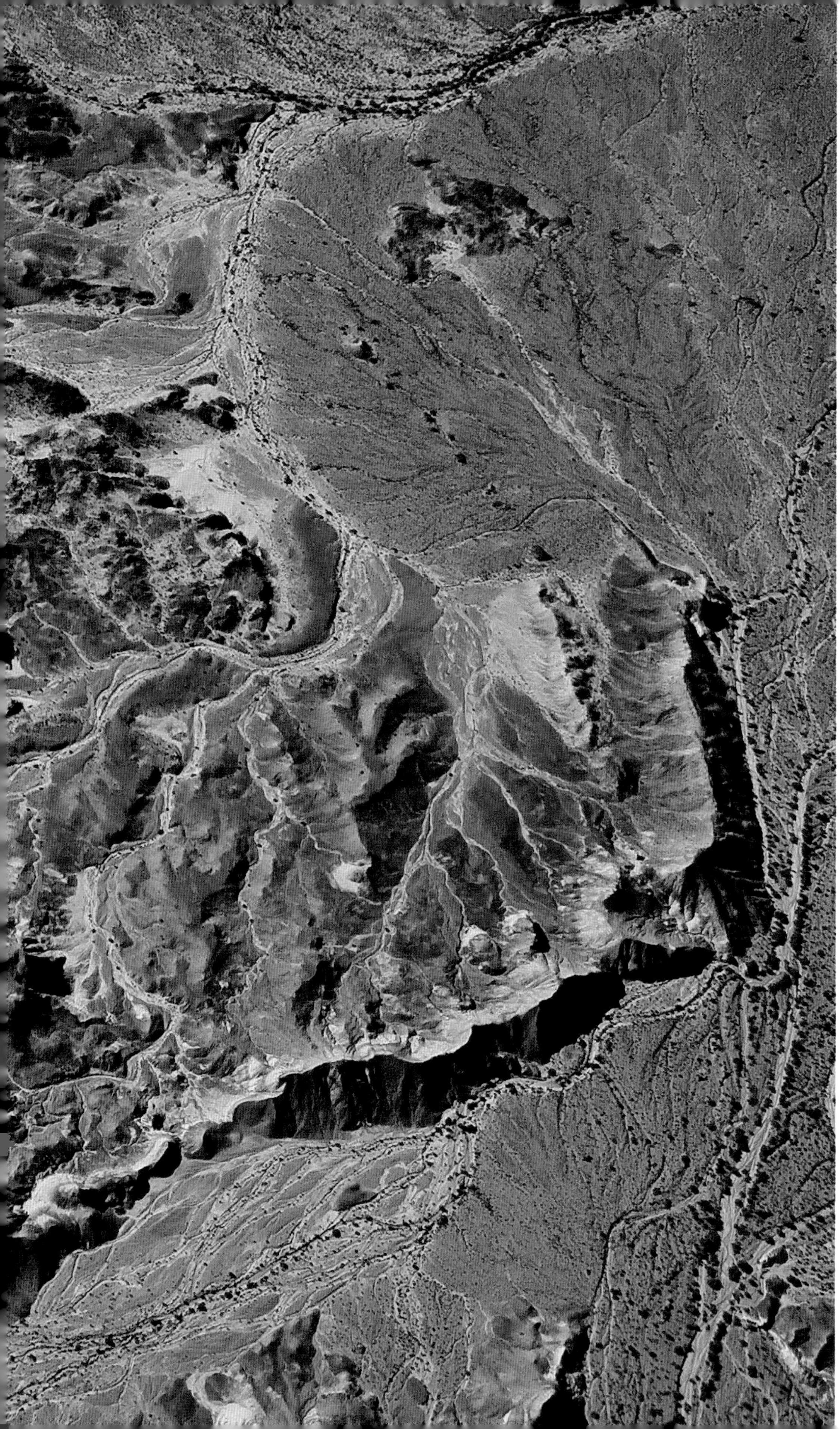

60. Erosion Flower

About five miles west of Mariscal Mine

The most dynamic agent in the geology of Big Bend today is erosion. In the summer, thunderstorms create flash floods with runoff that cuts the deep arroyos or ravines seen here—and then deposit the sediment on alluvial plains, locally called "flats." From the air, the result is a mysterious landscape with abstract organic shapes that beckoned to me as we flew over this amazing unnamed landform. To me, it looked like a flower made of sand, blossoming in the desert.

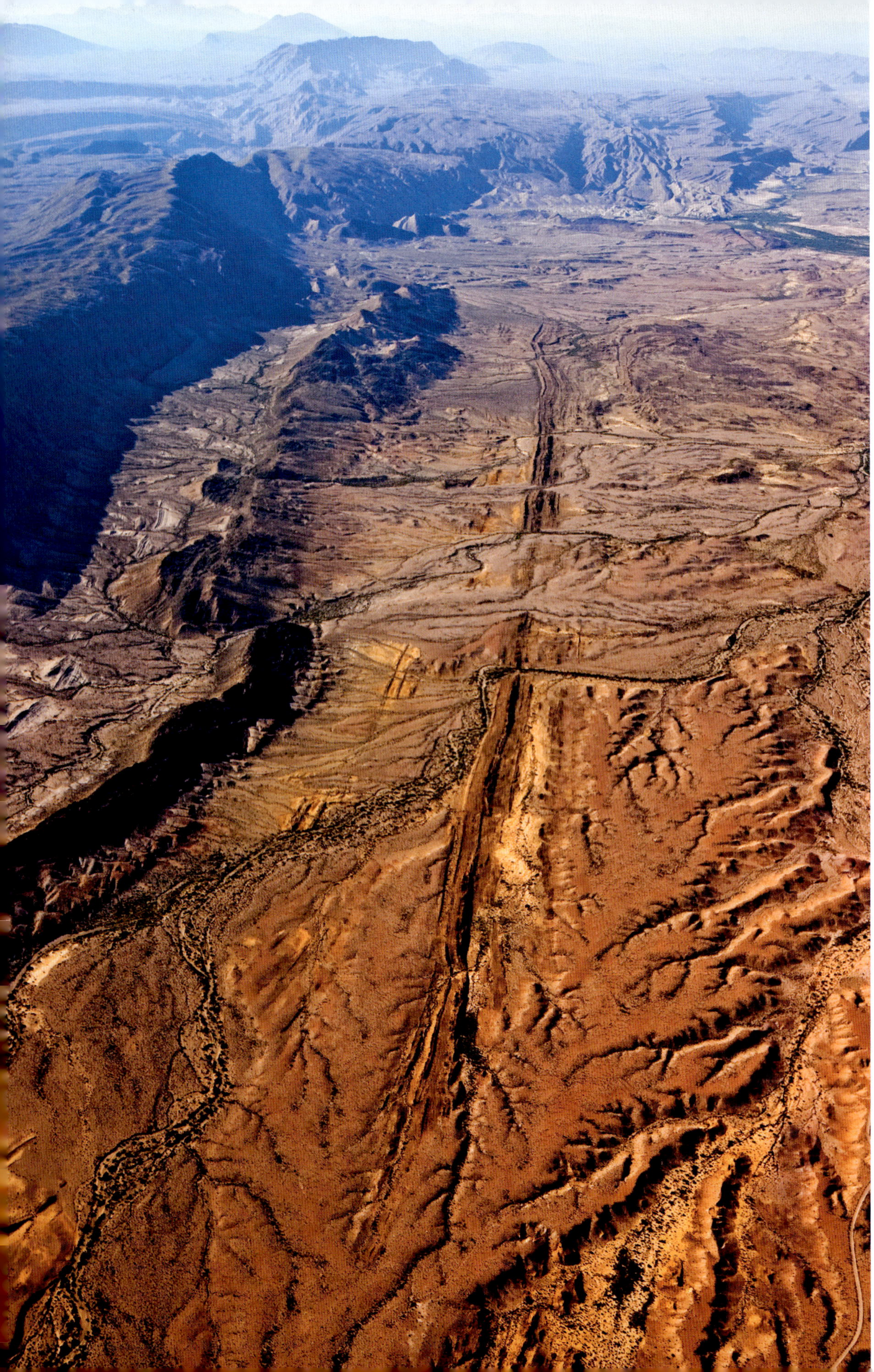

61. Earth Marks

Just west of Mariscal Mountain

Big Bend National Park is a "gold mine" for students of geology. On the west flank of the plunging anticline of Mariscal Mountain are other features such as normal faults, sills, and a syncline. In the distance the Rio Grande can be seen flowing right to left, entering Mariscal Canyon. Running down the center of this photograph is a mark on the earth that seems a beautiful straight line amid the tumult of the landscape. I have to remind myself that it is four miles long. In the distance, by the green landing at the upstream mouth of the canyon, is Talley (right), which takes its name from J. M. Talley, who ranched in Big Bend in the early 1900s. Talley is now the preferred place for primitive camping, putting in to the Rio Grande to float downstream ten miles through Mariscal Canyon to Solis, and as a trail-head for hikers exploring the exquisite but strenuous 3.3-mile Mariscal Canyon rim trail. High-clearance vehicles are a must to drive down the Talley Ranch Road to the river.

62. The Loop

Rio Grande

The green ribbon of life along the Rio Grande appears in startling contrast against the pale of the Chihuahuan Desert. On the other side of the Rio Grande can be seen Arroyo la Saladita in Mexico. In this U-shaped section of the river (left to right) are Loop Camp, Reed Camp, and in the distance Johnson Ranch. Elmo Johnson established a ranch, farm, and trading post here in 1927. Fur barter was the main form of transaction. In 1929, during the Escobar Rebellion, thirty bandits raided the outpost. In response, the US Army Air Corps built an airfield on the ranch to provide a means to rapidly send troops to the border. Johnson Airfield became a favorite destination for young airmen training in Texas. Here they found a place to practice their flying skills, go fishing, and enjoy exploring Big Bend. Some military training maneuvers were carefully planned to coincide with the opening of deer-hunting season. Big Bend aerial photographer W. D. Smithers worked from this base, where he creatively used military aviation resources. He even built an adobe darkroom using local construction methods on the Johnson Ranch, where, like my grandfather, he used the sun as a light source for making prints. The curving birdlike shape of the river shows how water has fashioned beauty in this arid wilderness.

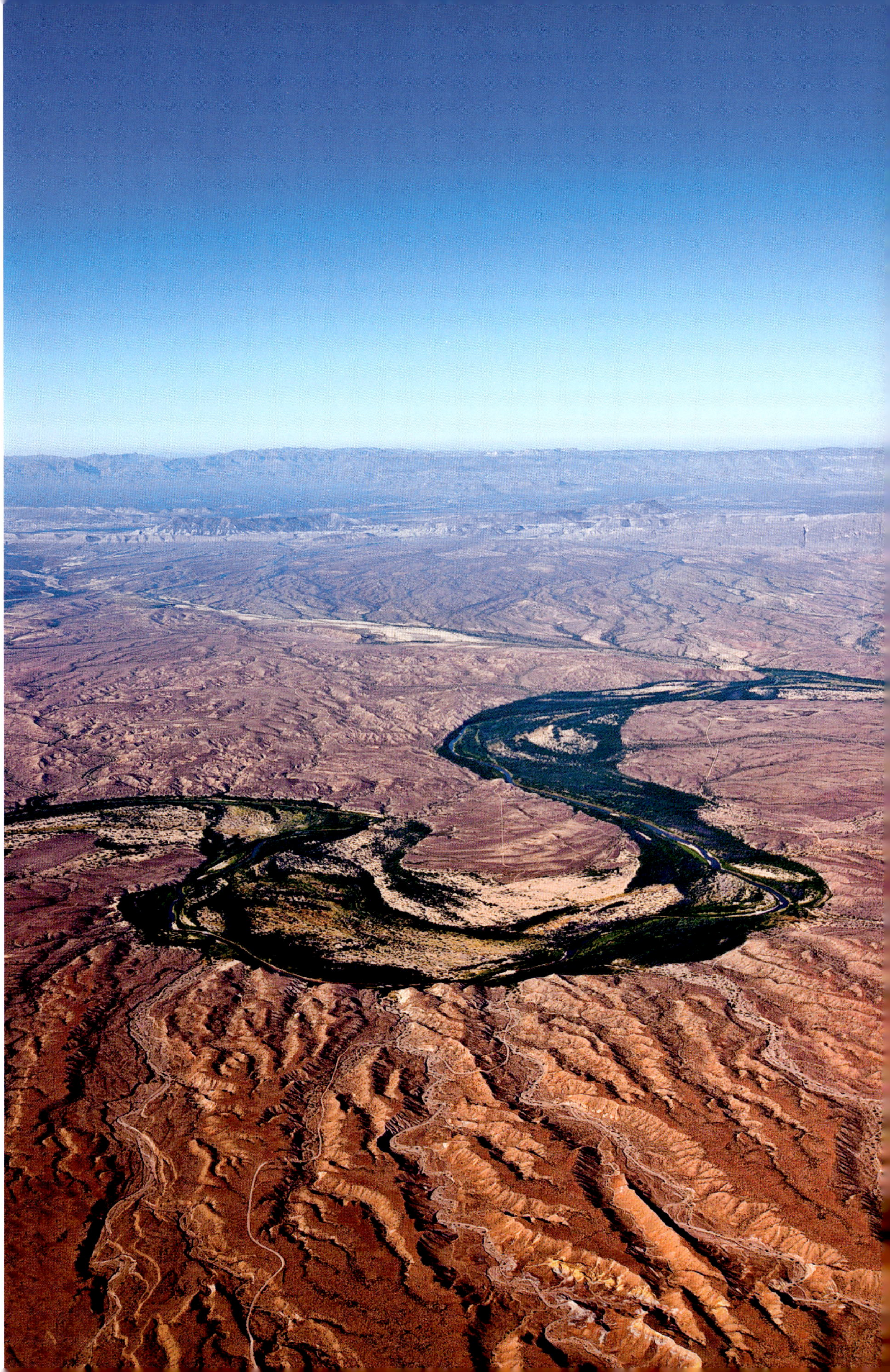

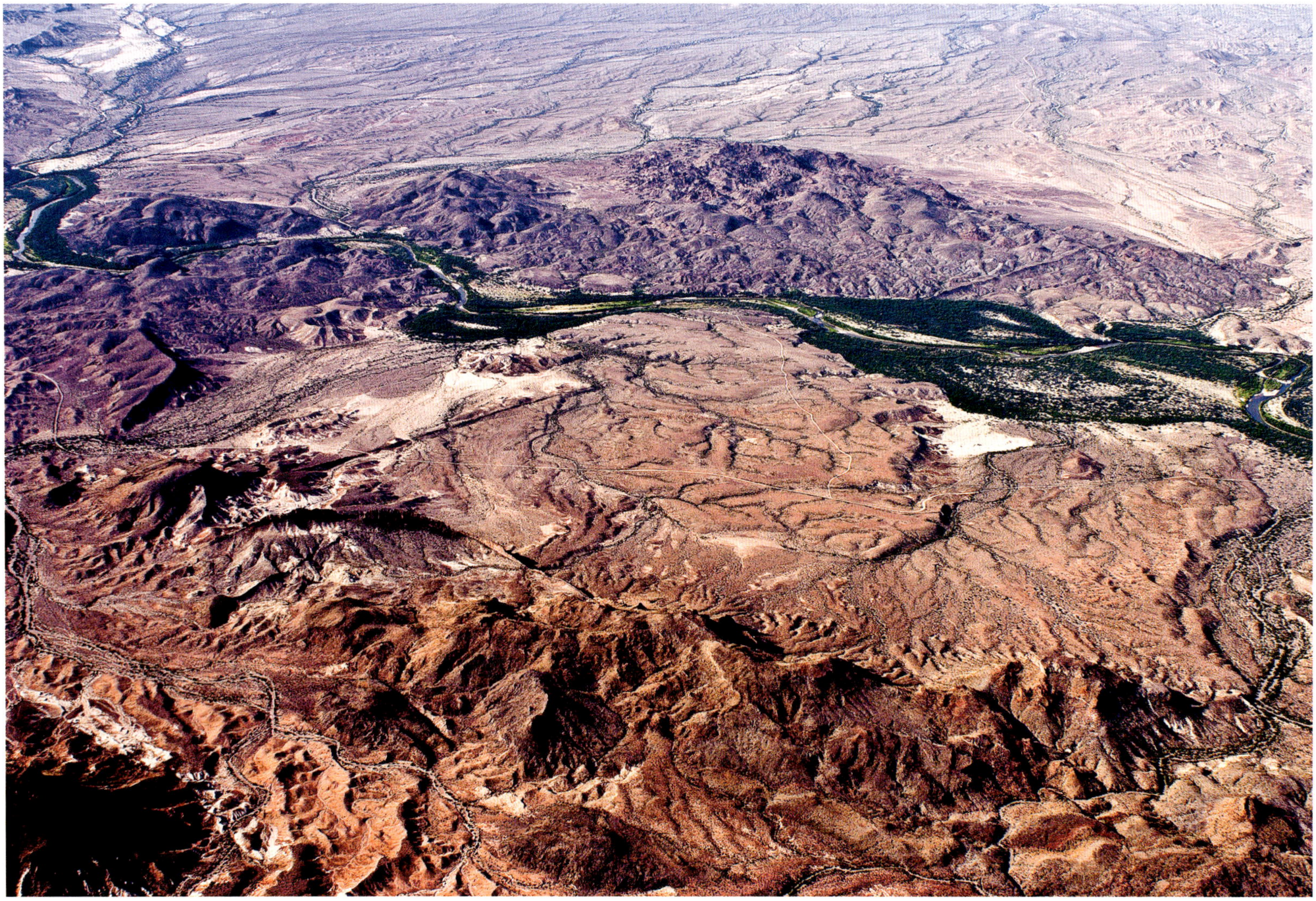

63. Sierra Chino Site

Downstream from Santa Elena Canyon

In the foreground are the Sierra de Chino. The road heading toward the river leads to the ruins and cemetery of the Sierra Chino settlement. Running left to right is River Road West. This area has several breathtaking ridges that look like the back spines of iguanas. They are geologic dikes formed when magma filled fissures, then cooled and hardened. Today only the igneous rock intrusions remain and can be seen as the dark diagonal lines just to the left of center.

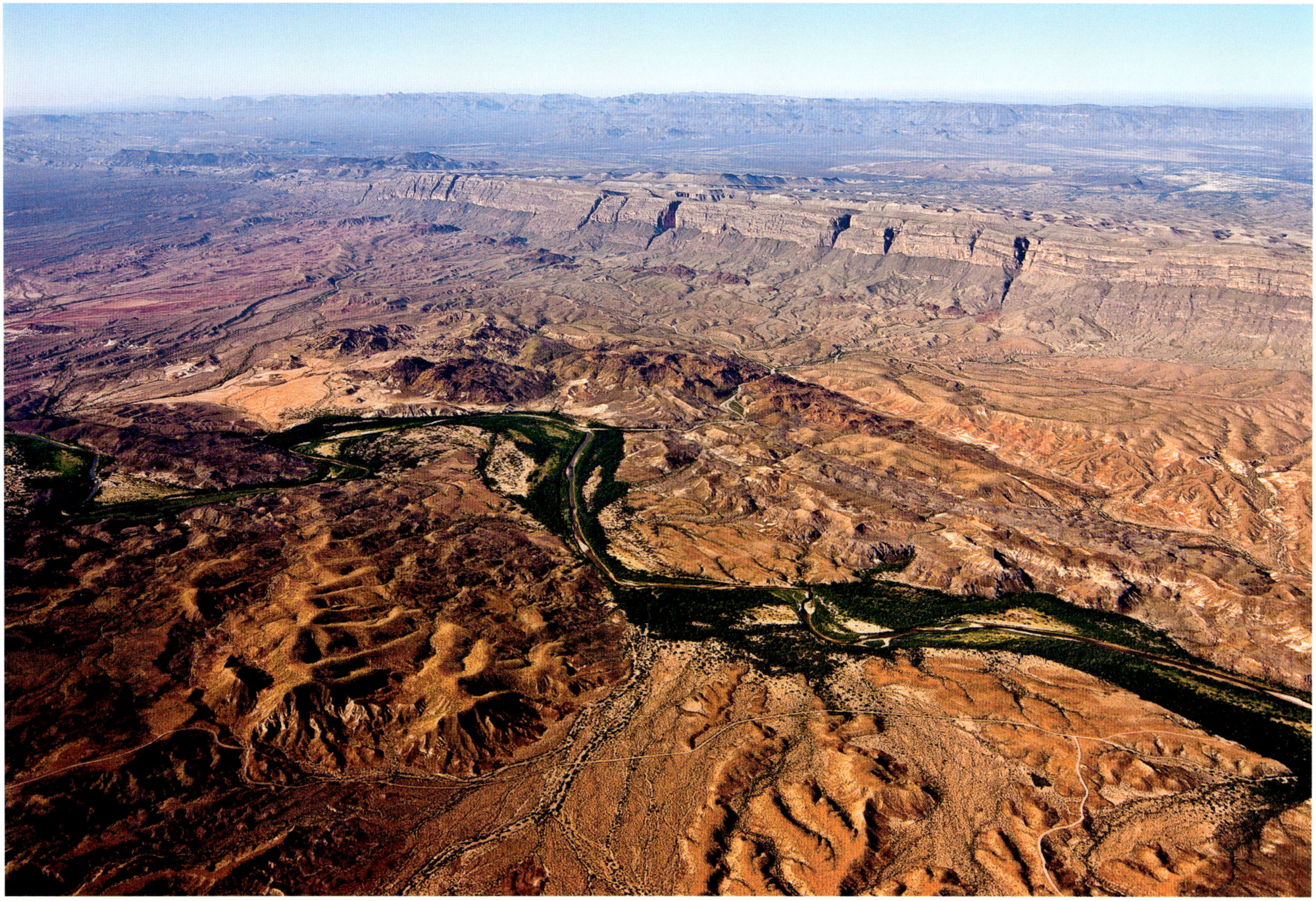

64. Sierra Ponce

Southwest of Castolon

Rising above the Rio Grande near Buenos Aires are the massive limestone cliffs of the Sierra Ponce of Mexico, with numerous unnamed canyons cut in their faces. Below the escarpment is Arroyo de Enmedio, cutting through Cerro Las Congojas (center) and meeting the Rio Grande at the Rancho Enmedio site. Along the Rio Grande on the US side (foreground) from left to right are the Smoky Creek site, Buenos Aires, and a scenic overlook circle on River Road West.

VI. CANYONS AND DRUG LORDS

The Rio Grande Upstream from Santa Elena Village to Lajitas

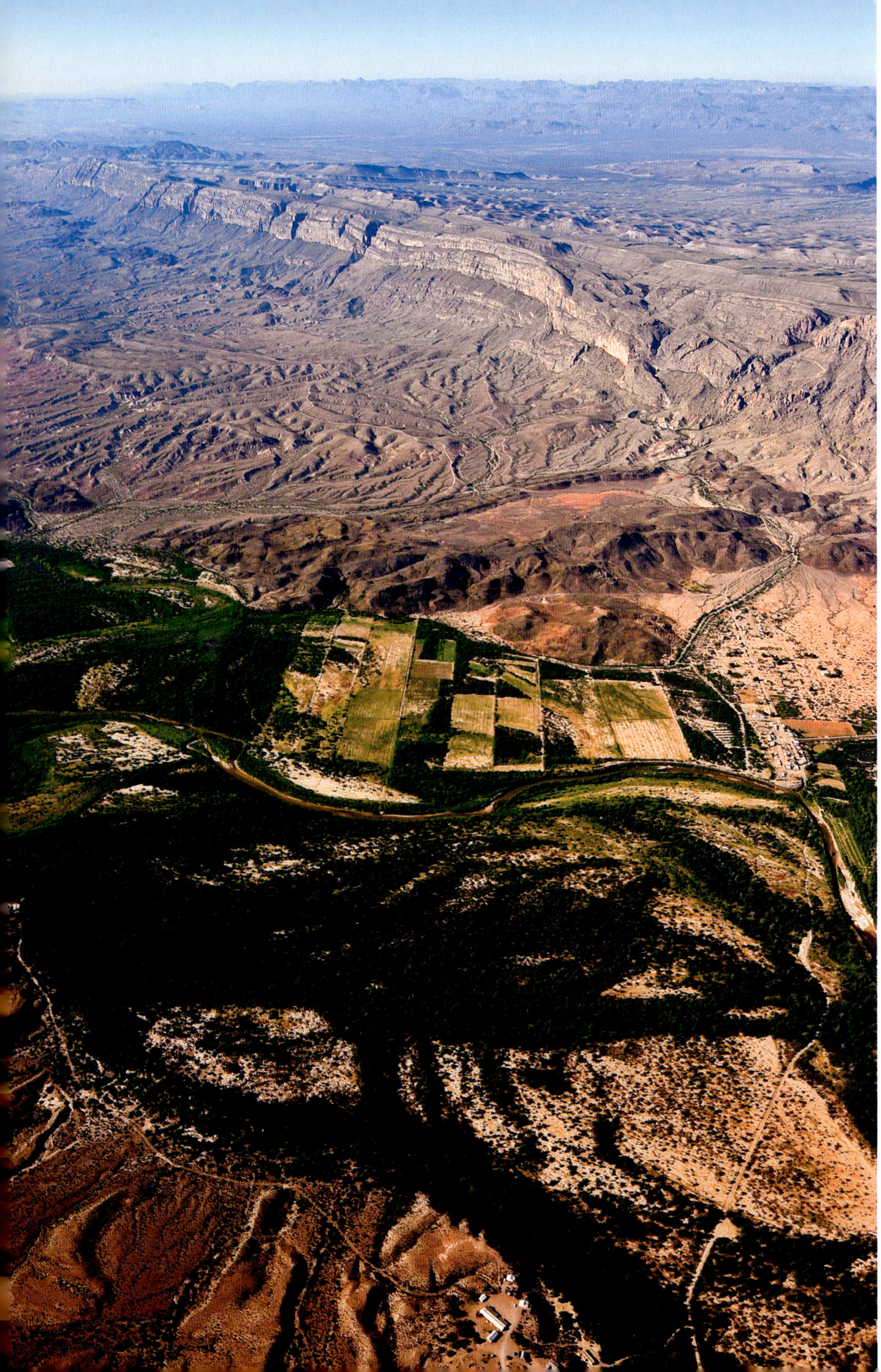

65. The Farmland of Santa Elena, Mexico

Just downstream from Santa Elena Canyon

Appearing like a verdant quilt in the desert across the river from Castolon, Texas (bottom, center), is the tiny village of Santa Elena, Mexico. Santa Elena was an *ejido,* or Mexican collective farm community. Families were relocated from less productive agricultural land to areas such as Santa Elena where irrigated farming along the Rio Bravo was possible. The *ejido* system that was critical to the survival of peasant and indigenous peoples was dealt final blows by the North American Free Trade Agreement, which sought to make the Mexican *ejido* lands available for purchase by foreign investors without fear of later confiscation. This peaceful-looking hamlet was also the place where Pablo Acosta, a top drug lord, or *padrino,* was born and later died in an hour-and-a-half-long machine gun battle with Mexican authorities. The Mexican federal police assault team took off from Panther Junction, then crossed the Rio Grande here in two helicopters in 1987. Acosta smuggled an estimated sixty tons of cocaine per year into the United States.

66. Rio Grande, River of Life

Above Castolon, downstream from Santa Elena Canyon

Just below its confluence with Terlingua Creek, the Rio Grande spreads its green fan across the Chihuahuan Desert as it flows out of Santa Elena Canyon. Winding to the right of the channel is River Road West. Above the river is Sierra Ponce (left). In the distance are mountains in the Mexican state of Chihuahua. Seeing the dramatic green band of life created by the waters of the Rio Grande motivated me to choose this photograph. I also found the camera angle intriguing in this rare view looking directly upstream, with Mexico on the left and the United States on the right. I believe that we were at least a few feet inside United States airspace, thanks to a turn in the river below us.

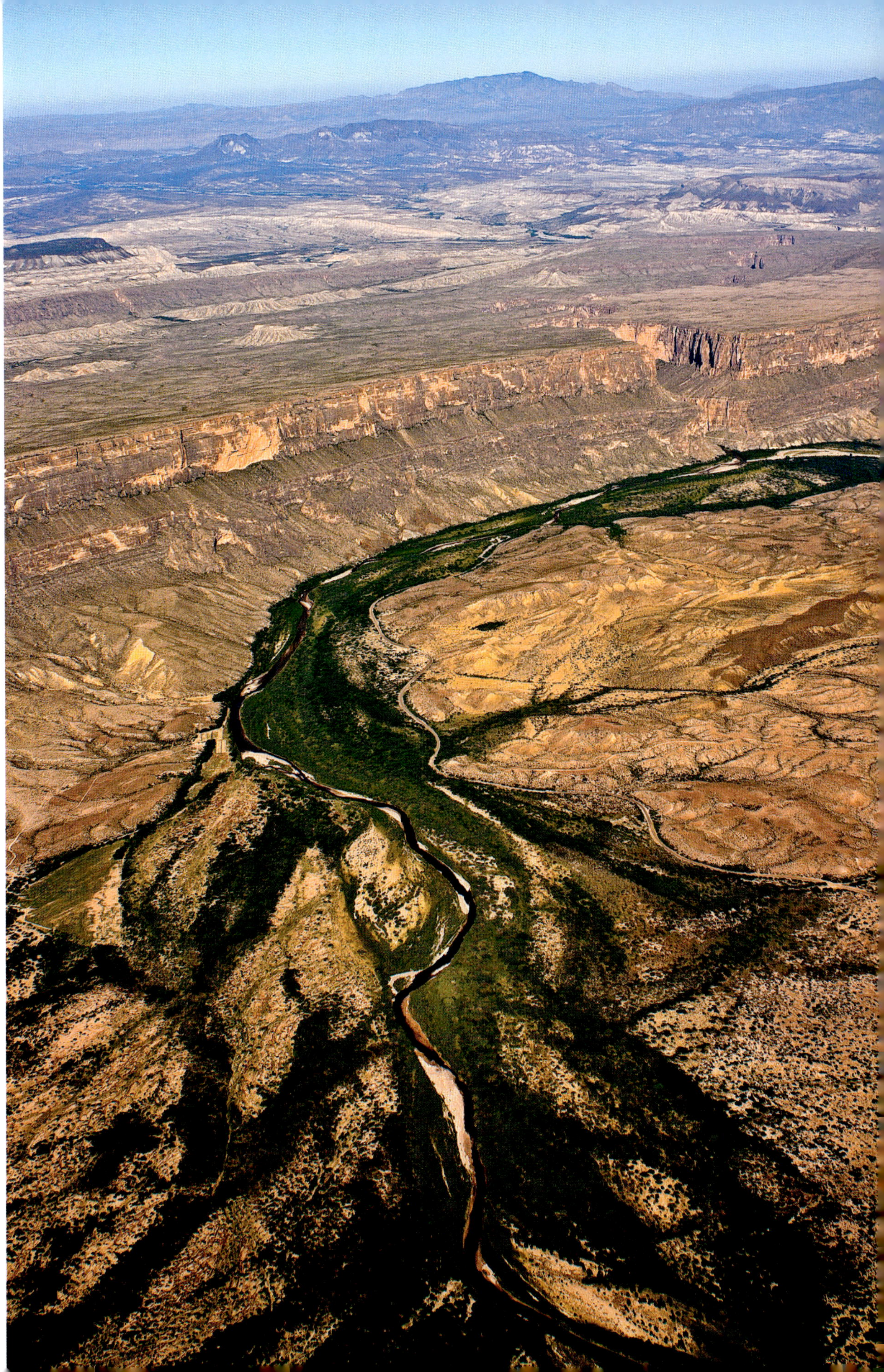

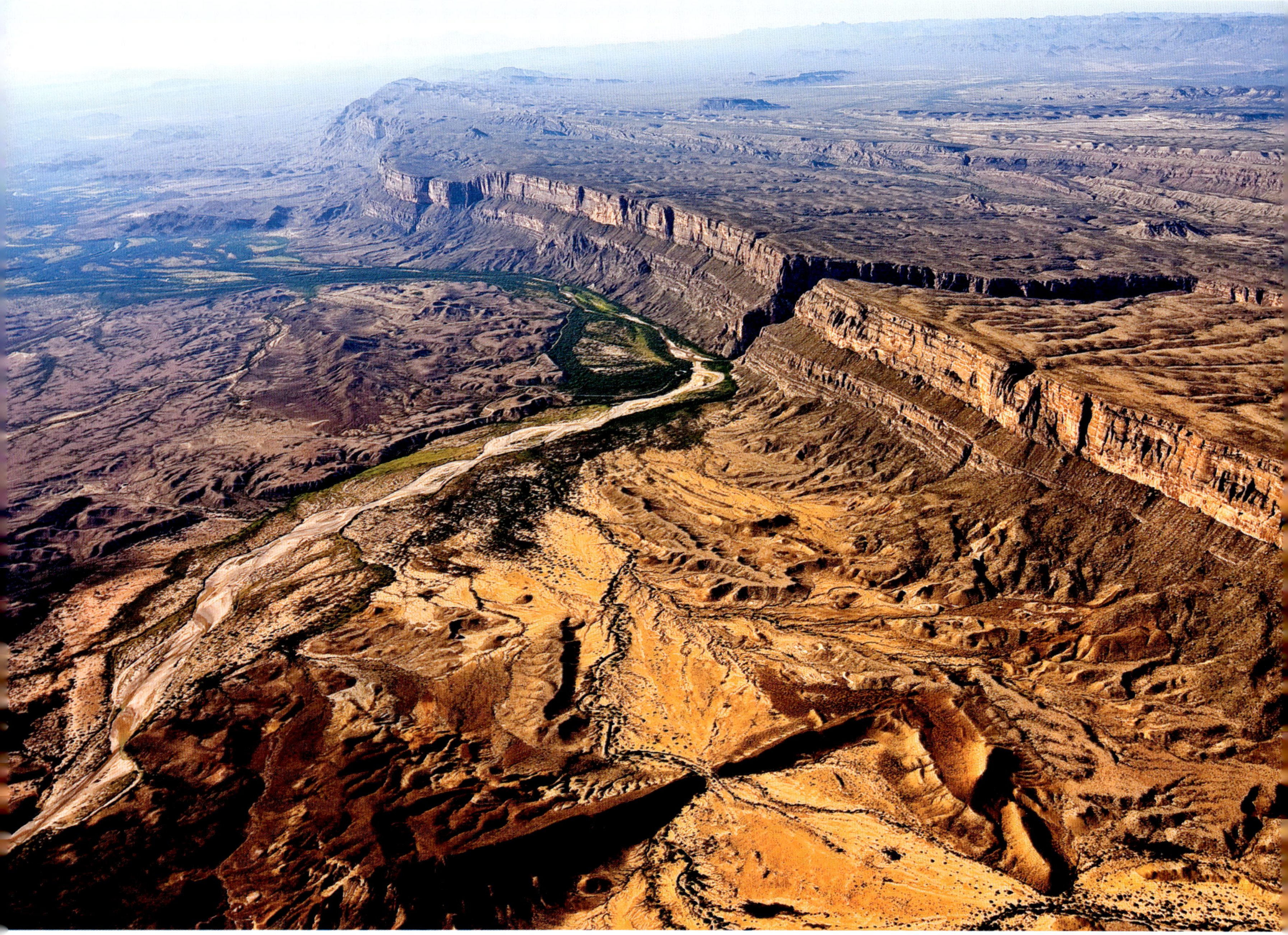

67. Santa Elena Canyon

Mesa de Anguila and Sierra Ponce

Here the Rio Grande emerges from Santa Elena Canyon (right). The canyon, about 1,500 feet deep, cuts through the Boquillas limestone of the impressive Mesa de Anguila, which was created by uplift along the Terlingua Fault. On the Mexican side of the canyon (top), the mesa is called Sierra Ponce. This view shows about sixteen miles of the rock face. In the lower part of the image (left to right) are Terlingua Creek and a magmatic dike cut by an arroyo. I like how this view shows the Santa Elena Canyon downstream mouth, a popular tourist destination, in a fresh way, pitted against the formidable mesa and the stunning river valley.

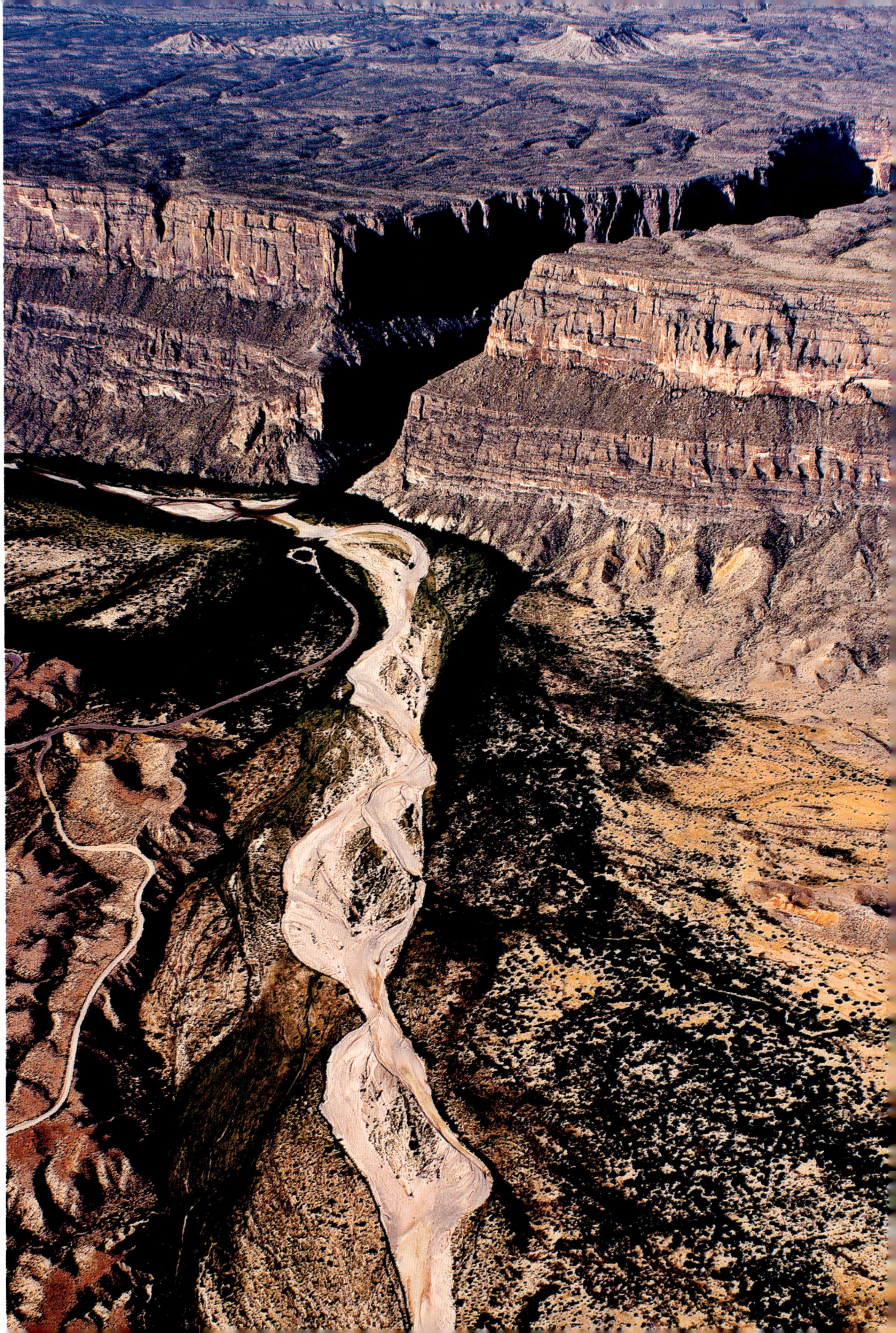

68. Overlooking the Scenic Overlook

Santa Elena Canyon

As we flew toward Santa Elena Canyon, I remembered how Ansel Adams photographed this canyon from the ground in 1947. Here the Rio Grande emerges from the steep face of Mesa de Anguila and is met by Terlingua Creek (center). Our aircraft climbed higher in altitude as we prepared to pass over the edge of the escarpment and travel toward the upper mouth of the canyon. The hard edge of the canyon wall separates the dark inner passage from the blinding light of the desert. I chose this portrait orientation, rather than a horizontal view, to emphasize the sheer vertical walls of Santa Elena Canyon.

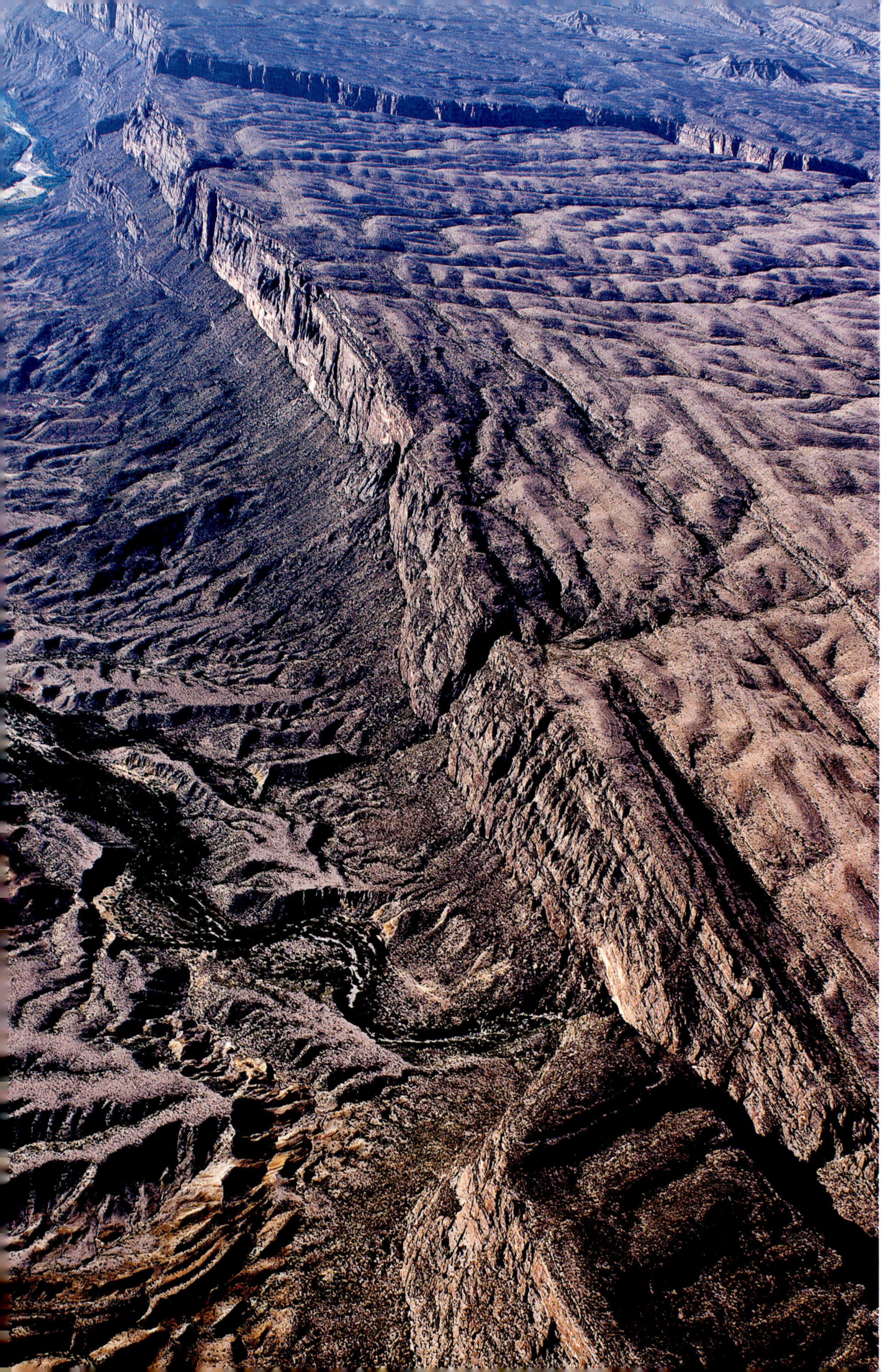

69. Crossing the Terlingua Fault
North of Santa Elena Canyon
The edge of Mesa de Anguila, with Santa Elena Canyon in the distance. Below is Bruja Canyon (lower right corner); *bruja* is the Spanish word for a Mexican spirit-healer. A hiking trail rises from the desert floor through Bruja Canyon as a shortcut to the top of the mesa, but the route is dangerous and requires rock-climbing expertise.

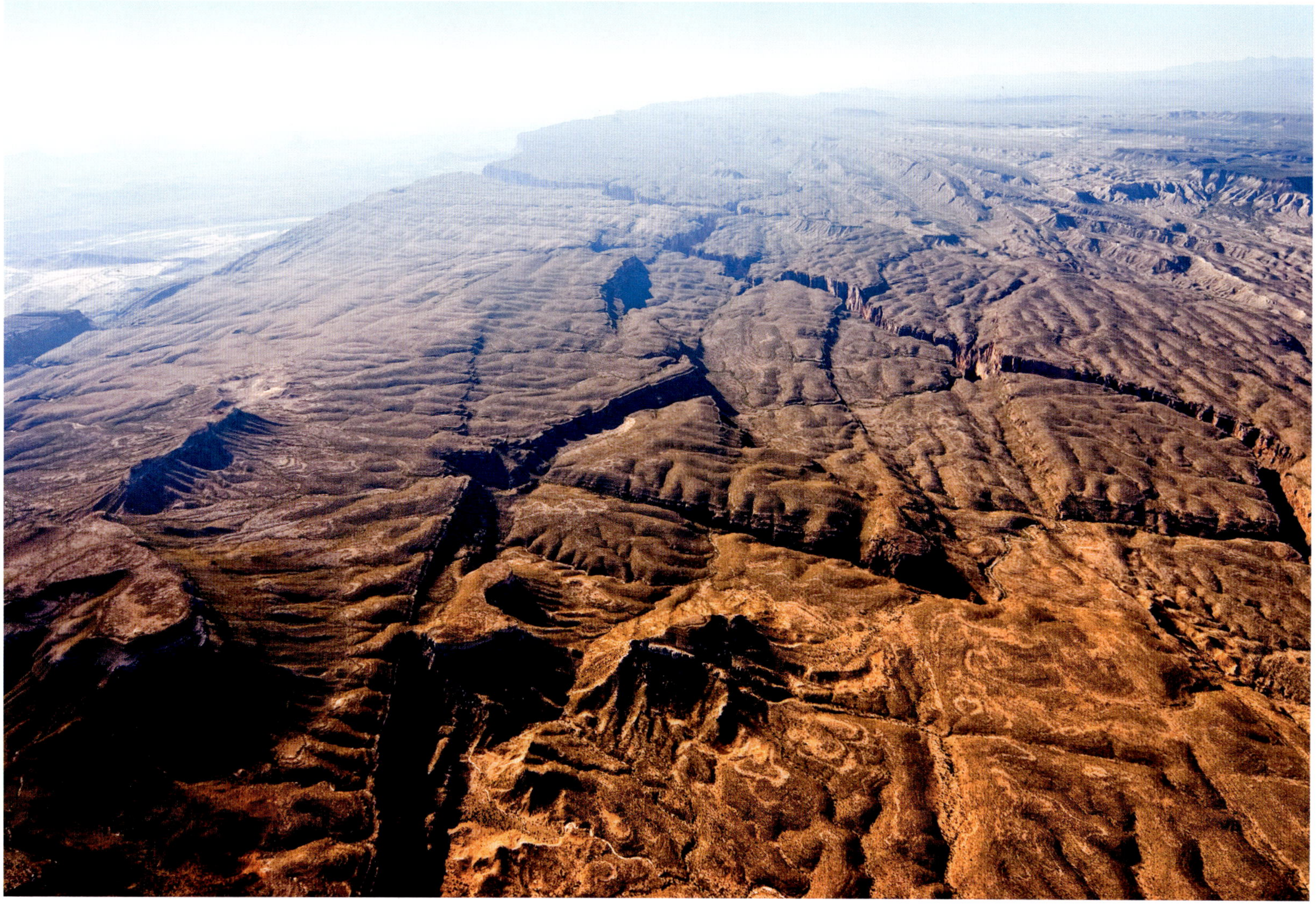

70. Top of Mesa de Anguila

Looking south into Mexico

The top of Mesa de Anguila swelters in the summer sun, yet is often visited by hikers who seek views of Santa Elena Canyon from the rim above. Canyon Flag (foreground, left), elevation 3,883 feet, is the highest point of the mesa. Below that is a pass with a trail that leads between the ephemeral water holes of Tinaja Lujan and Tinaja Blanca. I made this photographic view wide to seize the moment when I first saw this seemingly endless expanse of the mesa top, so gargantuan in scale that the huge Santa Elena Canyon below appears lost amid the blanket of texture. This is an almost unimaginable perspective to a paddler in the river or a hiker on the ground.

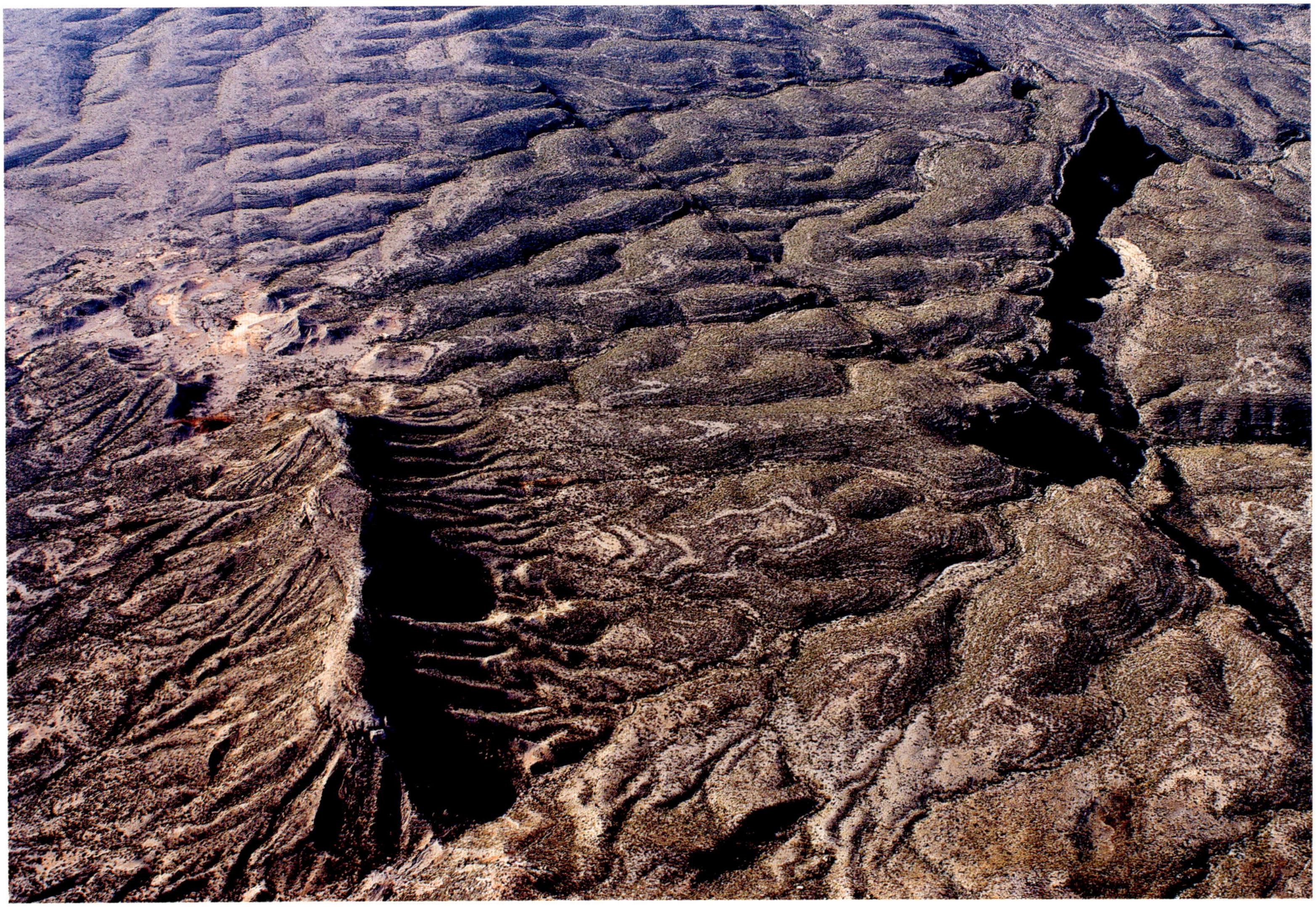

71. Magma Dike on Mesa de Anguila

Just south of Canyon Flag

In this detail of the top of Mesa de Anguila, the geologic forces of sedimentation, faulting, magma intrusion, and erosion combine to form an alien-looking landscape. The visually striking ridge (left) was formed by molten rock pushing into a crack in the limestone layer and forming a dike. To the right is a scalloped ridge and depression in the mesa that collects the sparse rainfall and temporarily holds the water in natural rock pools known as *tinajas.* I carefully framed this shot while speeding across the mesa top to show intuitively the fascinating visual counterpoint of the mountain at left "playing off" the tinaja at right, with this raw open space between them.

72. Mesas on Mesa

Just south of Tinaja Blanca

Above Santa Elena Canyon is this miniature unnamed canyon (right) formed by the runoff of passing summer thunderstorms on the plateau. As we flew at low altitude over Mesa de Anguila, the cross shape of this landform reminded me of the shadow of an aircraft or bird on the arid mesa below. Another eye-catching feature is the "mesas on top of mesas" effect as seen in the lower part of this photograph.

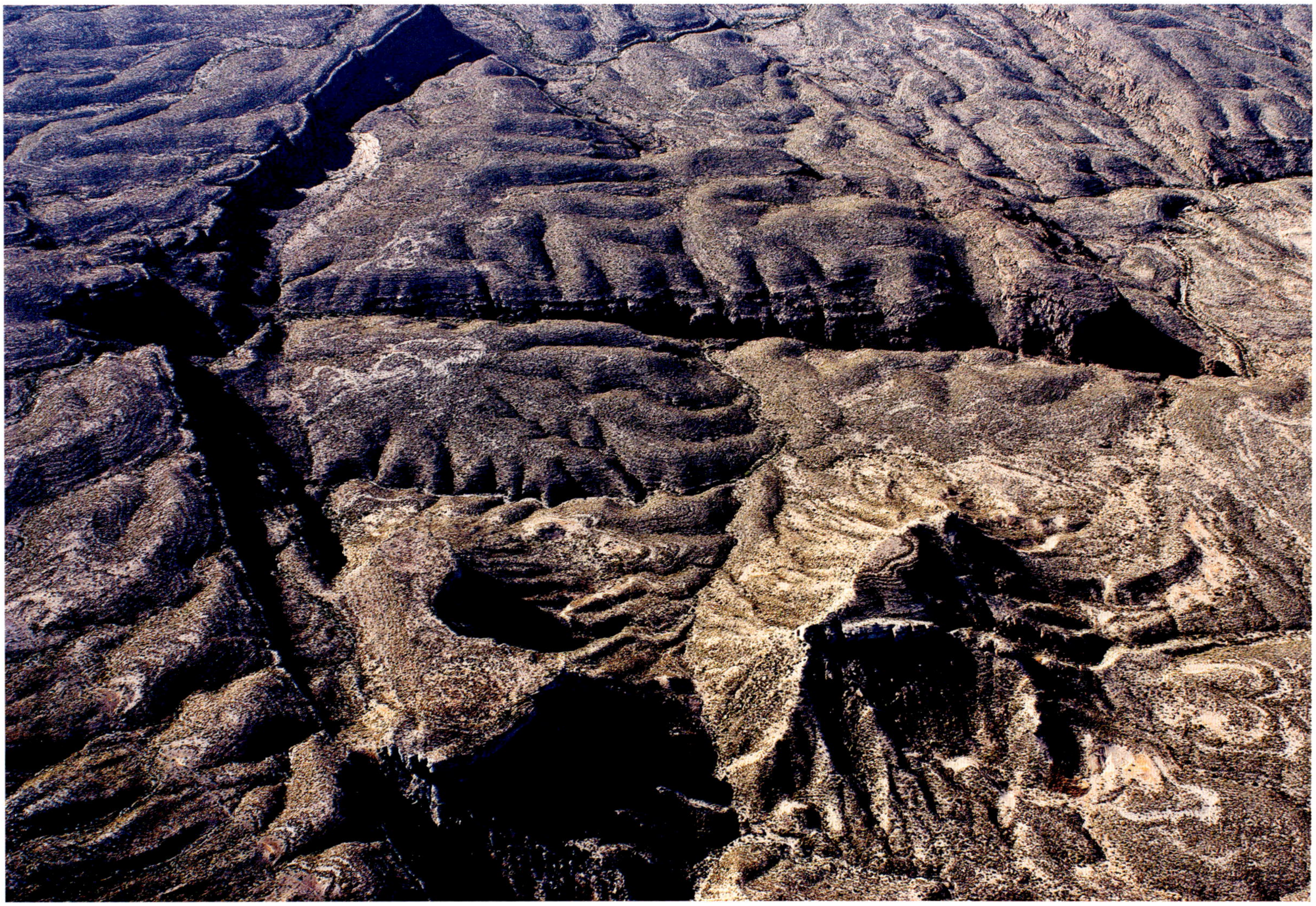

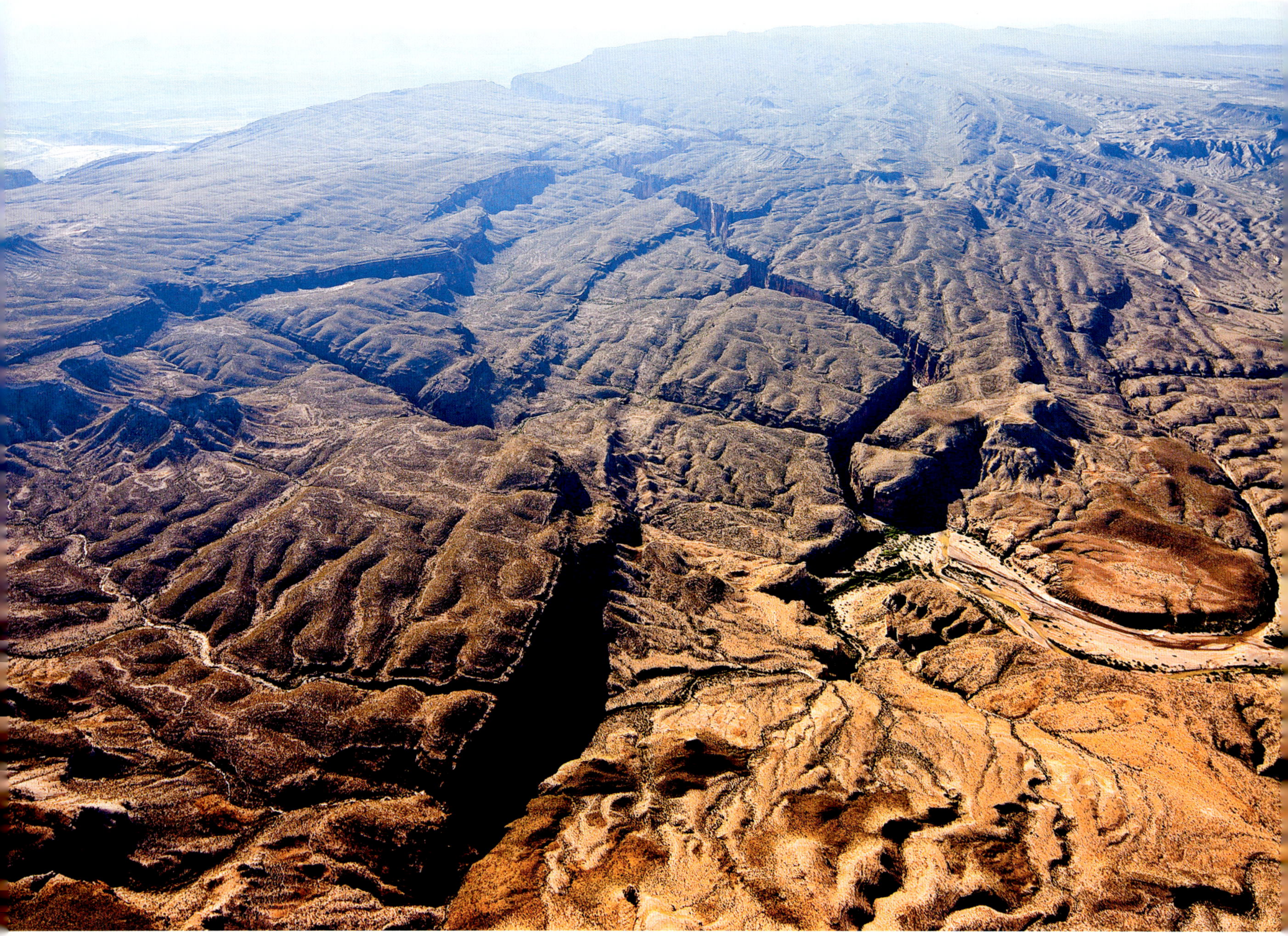

73. Mesa de Anguila and Sierra Ponce Vista

Downstream from Lajitas

From right to left along the Rio Grande are the confluence of San Carlos Creek, Entrance Camp, and the full length of Santa Elena Canyon. About a hundred yards inside the canyon are the Mirror Twins, a pair of canyons that bisect Santa Elena Canyon. Two miles downstream is Rock Slide, an area of truck-size boulders strewn in the river, which turns the Rio Grande into a Class IV rapid at higher water levels. I was spellbound by the fantastic marks that erosion scribed into this extensive mesa.

74. Upstream Mouth

Santa Elena Canyon

Following the Rio Grande from the bottom of the photograph upward are the Doldrums, San Carlos Creek, and Santa Elena Canyon. The straight stretch of the Rio Grande is know as the Doldrums (lower right), as there is little current and paddlers must work hard to make headway before reaching the swift water of the canyon. The sandy landing (left of center) at Entrance Camp is a favorite spot for float-trip paddlers to stop for the night before continuing their journey into the spectacular 1,500-foot depths of Santa Elena Canyon. In making this photograph, I sought to capture the whimsical shapes of the river while being cognizant of the tremendous forces and eons of time that it took to form this scene.

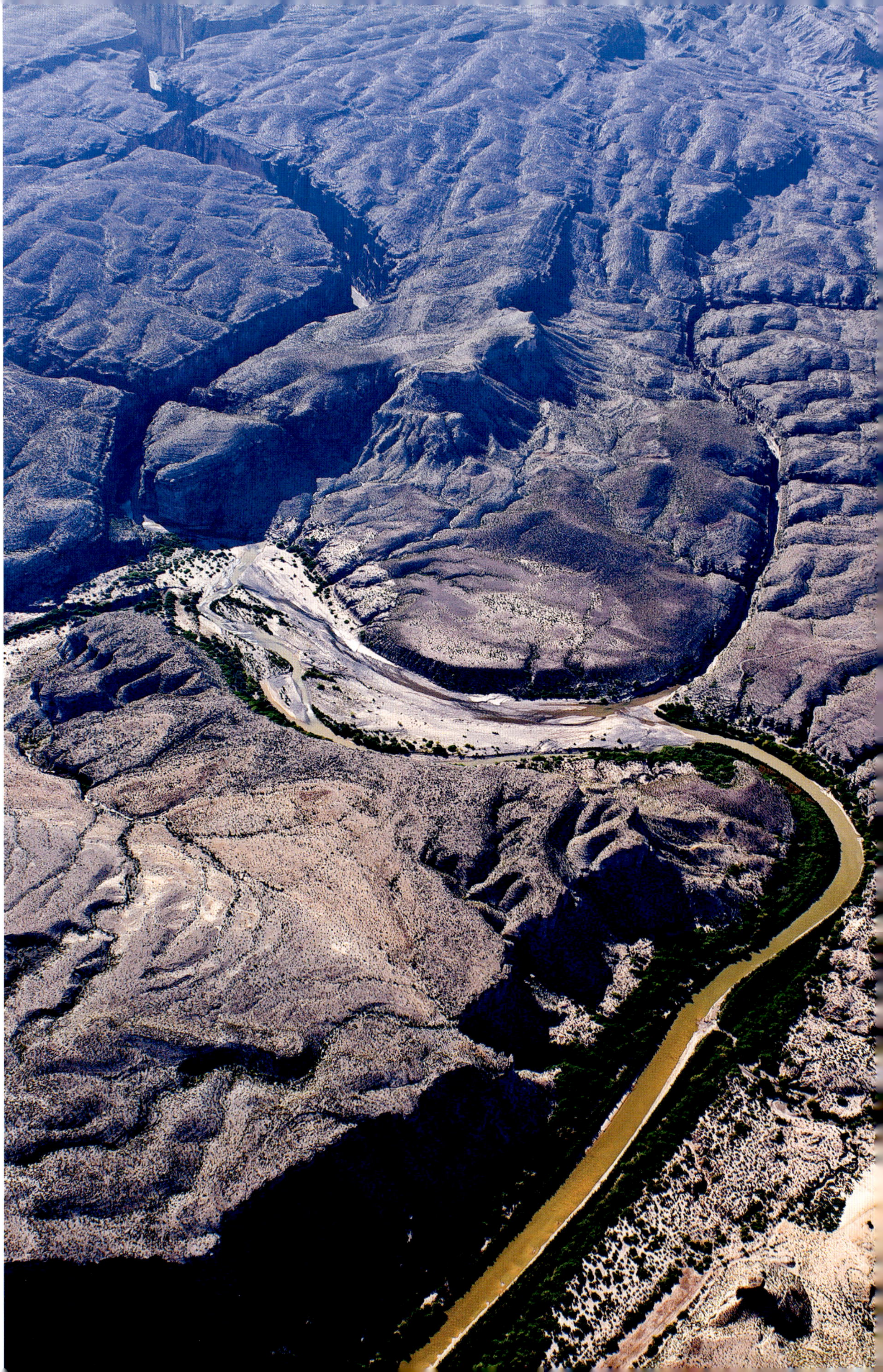

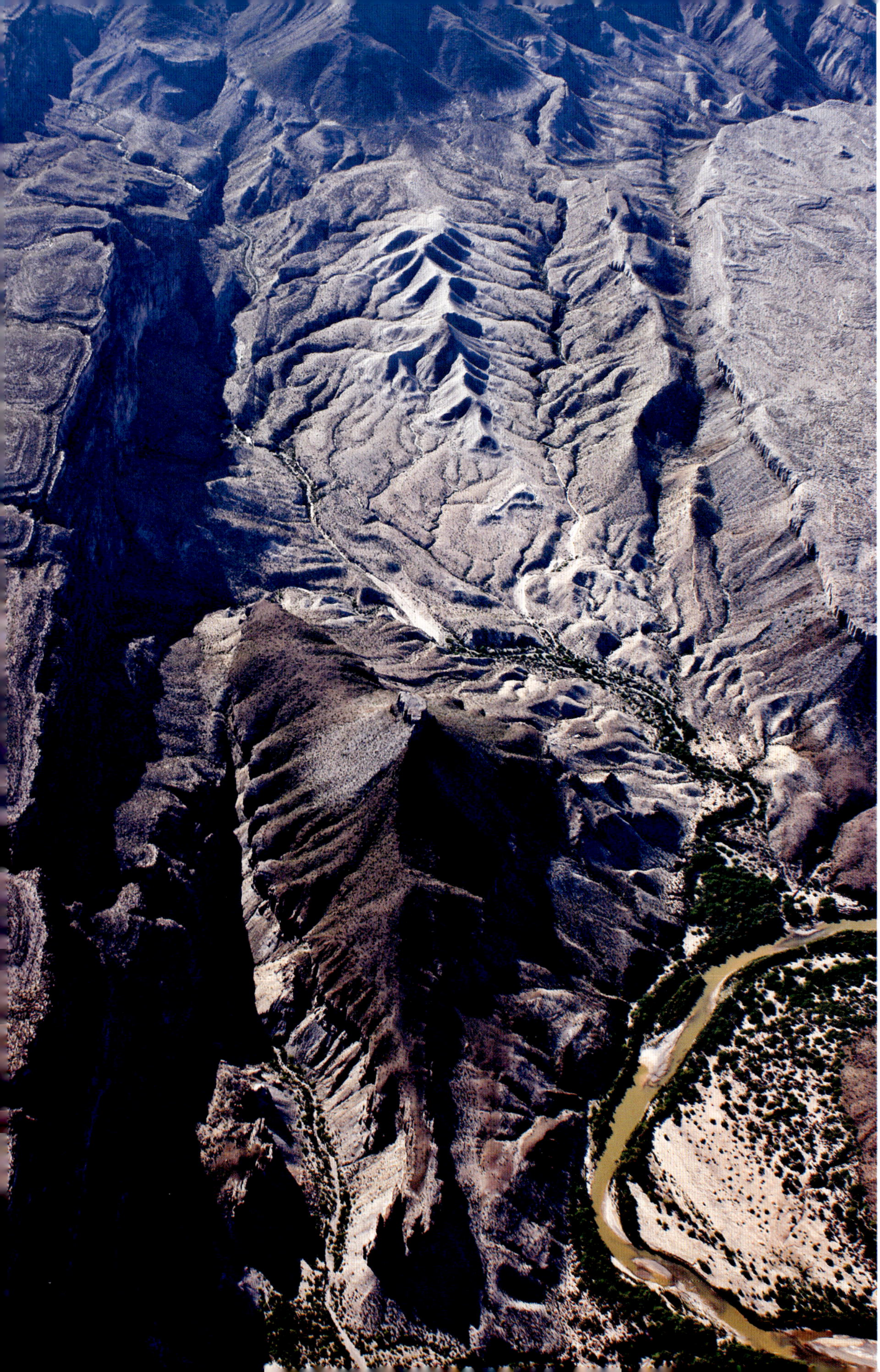

75. Metates Camp

On the Rio Grande

Metates Camp (right) takes its name from the numerous *metates,* or mortar holes, eighteen to twenty-four inches deep in the rock where indigenous people ground mesquite beans in preparation for cooking, much like grinding corn for tortillas. The adjacent mountain, False Sentinel (right), is often mistaken for the true Sentinel Peak that marks the entry to Santa Elena Canyon. River travelers who make this navigation error find themselves much farther upstream than the expected waypoint. A trail from Mesa de Anguila near Tinaja Blanca descends to the river here for hikers who take the overland route from the Saddle or Bruja Canyon.

76. Rio Grande

Above Santa Elena Canyon

The area downstream from Lajitas resembles a Cubist landscape with disjointed views of canyons, hills, and river that stirred my artistic imagination. Many recreational boaters in rafts, canoes, and kayaks make their journey down the Rio Grande through this remote area. They typically put in at Lajitas, float downstream through Santa Elena Canyon, then exit the river. The eighteen-mile trip typically takes two days. Among other river-use regulations, paddlers are required to obtain a National Park Service back-country use permit, wear a life vest, and carry a spare paddle in case one is lost in the rapids. River conditions vary widely from sparse water during dry spells to raging whitewater rapids following rainstorms.

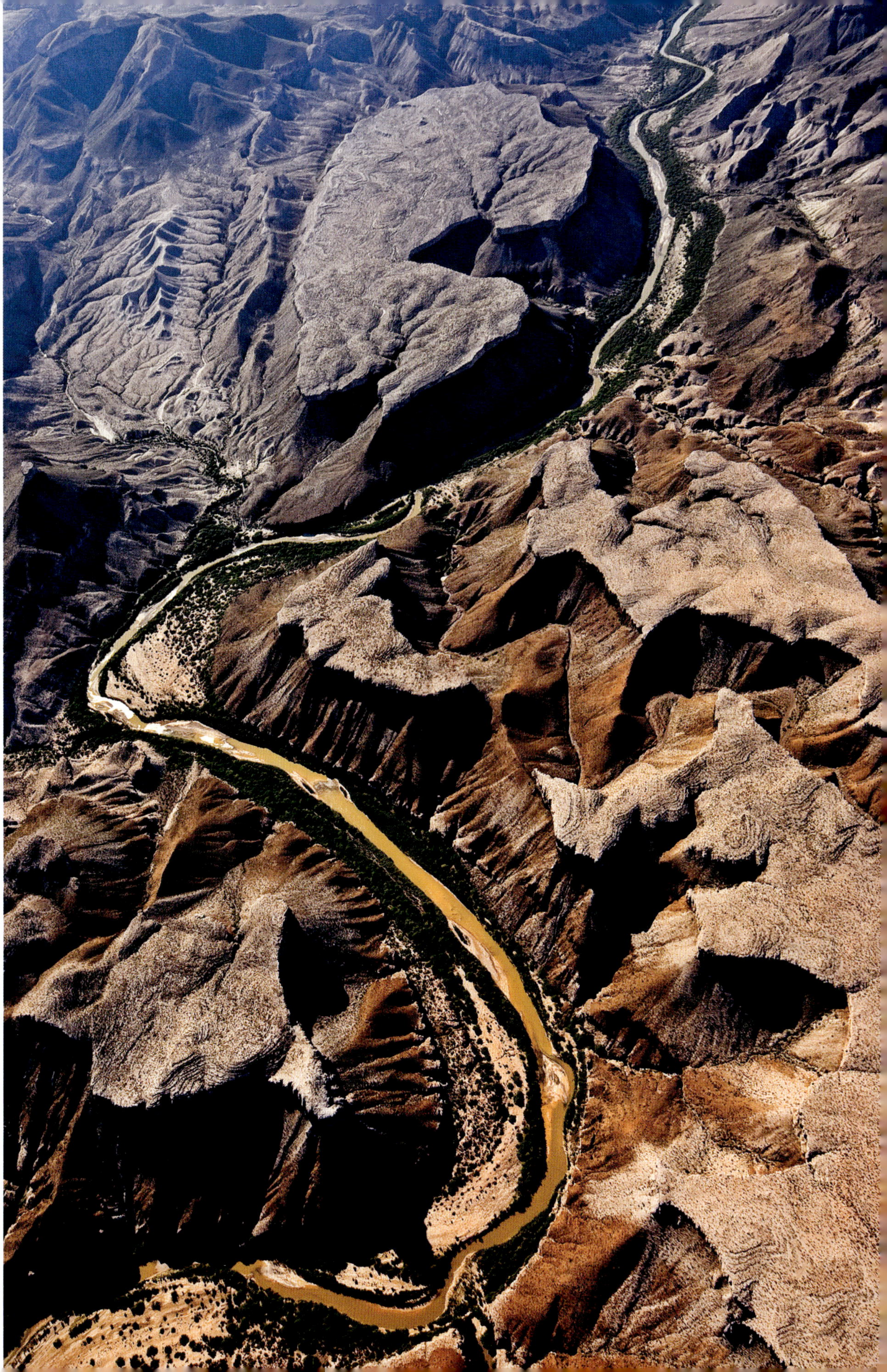

77, 78. Tuff Canyons (1 and 2)

Just below Lajitas

In the foreground (below) a canyon is carved into a massive layer of tuff by the Rio Grande. The tuff was created during a pyroclastic volcanic eruption that showered the area with glowing ash that settled and cooled under the pressure of its own weight. Tuff is easily eroded, in contrast to the more resistant limestone of Mesa de Anguila. Paddlers refer to the strange landform (below left) as King Kong for its gorilla-like silhouette when viewed from the river. Rather than using one photograph, I decided to present this view in two photographs shown side by side to help give a sense of the grand scale of the terrain forged by volcano and water.

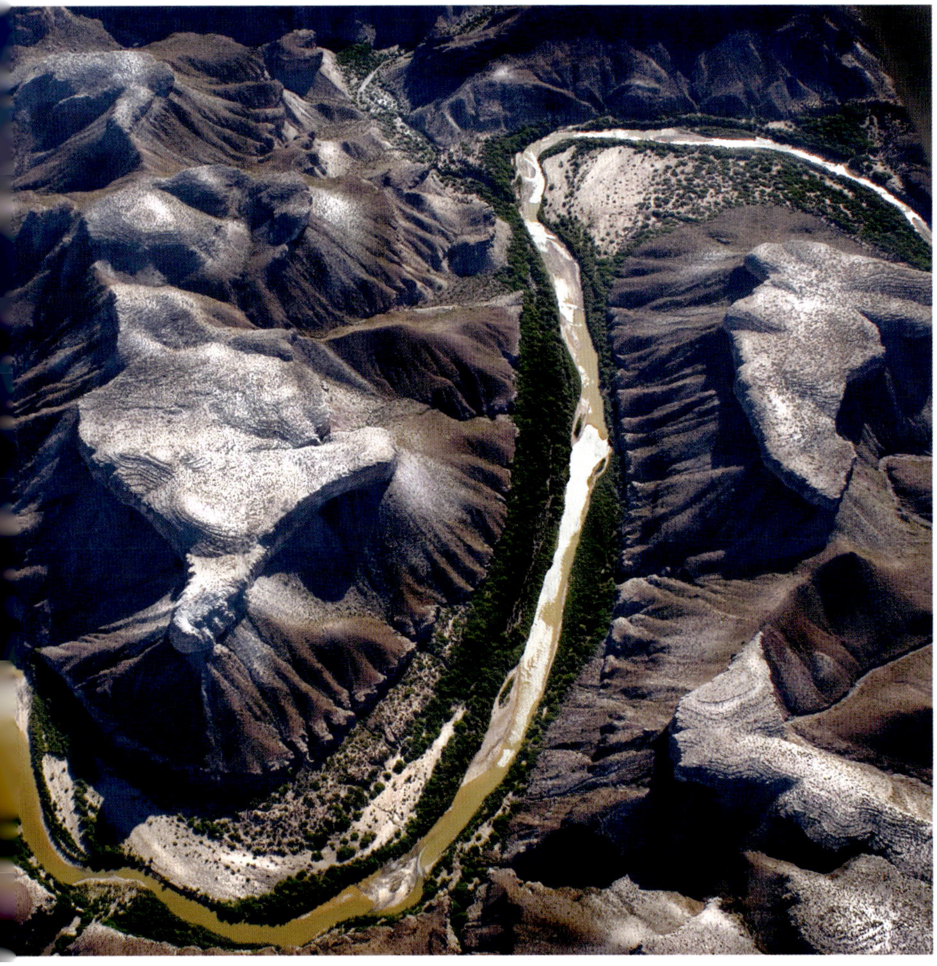

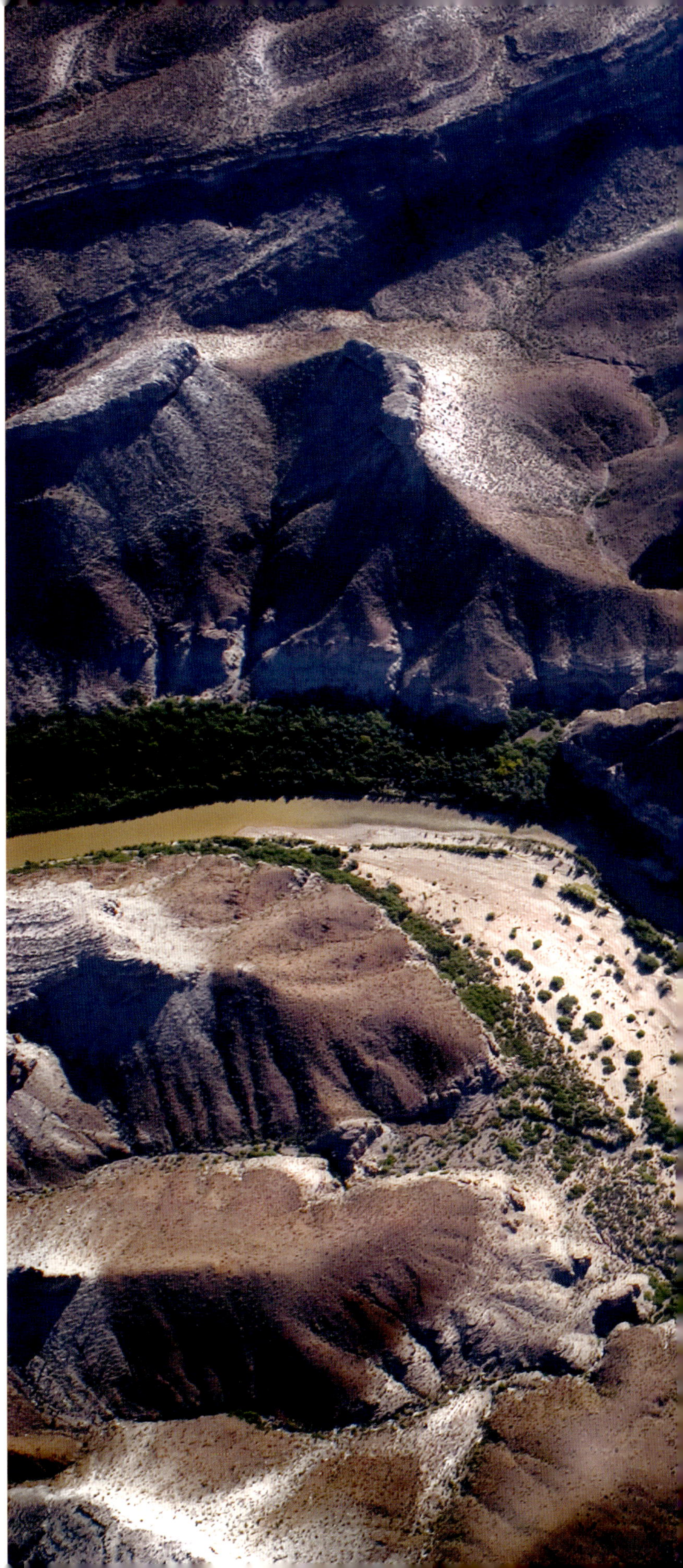

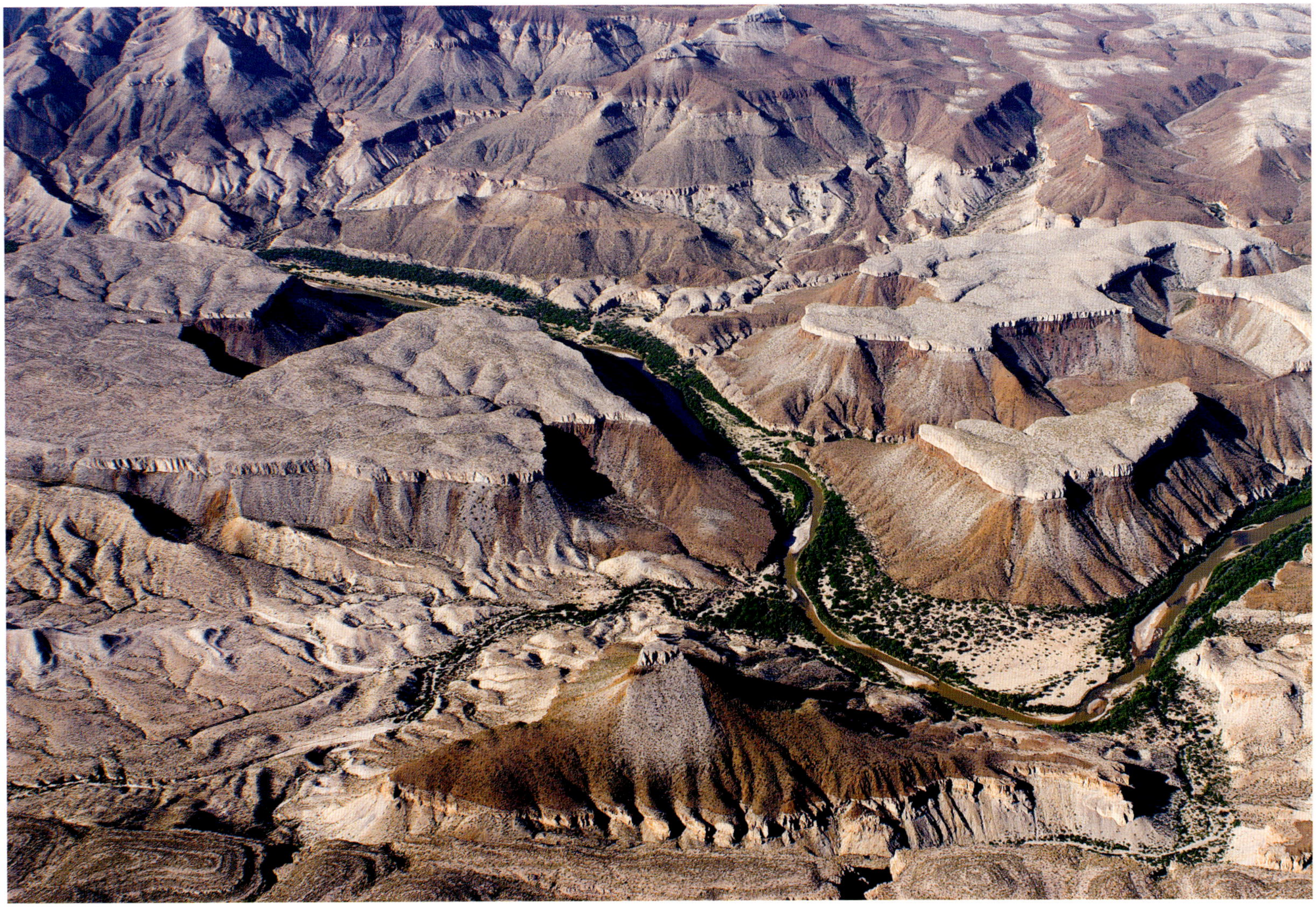

79. False Sentinel

Looking west into Mexico

The Rio Grande was originally named Río Bravo, or "wild river," by Spanish explorers and is still called that on the Mexican side of the border. The mountains just across the river are the Sierra Los Encinos (upper left). In the foreground, the tuff mesas are topped with limestone caps, creating a look similar to the crust on a cherry cobbler. I asked my pilot to fly lower for this more oblique or side view to reveal the colorful slopes of the tuff hills.

80. The First Good One

Matadero Canyon

To river travelers who have endured the monotony of the Rio Grande from Lajitas to this point, the confluence with Arroyo del Matadero (upper right) marks the first whitewater, locally known as the First Good One. *Matadero* means "place of slaughter" in Spanish. In the shadow of Matadero Canyon, pine trees line the river where swift current begins. The knife-edged mountain (left of center) is part of the Sierra Los Encinos in Mexico.

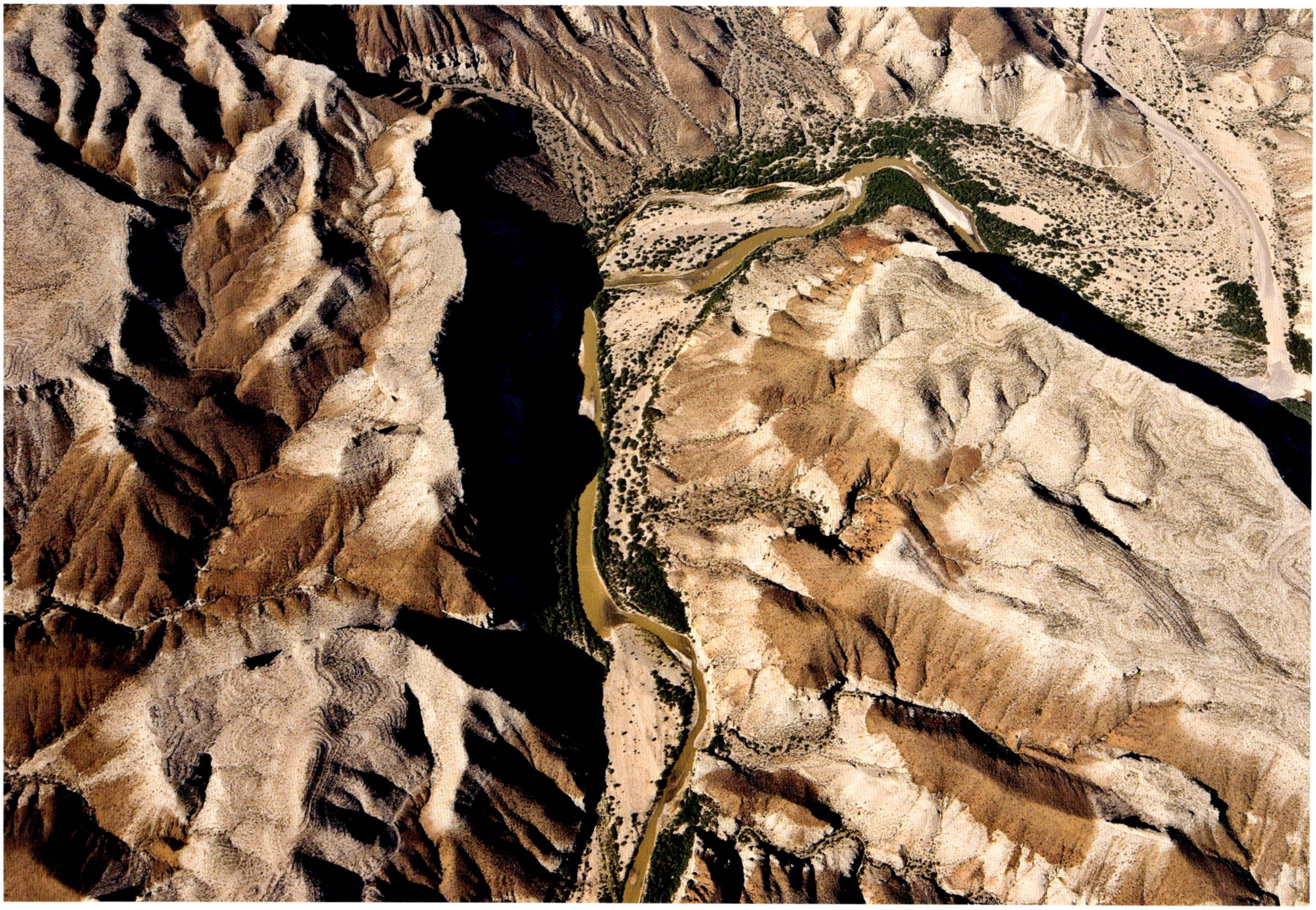

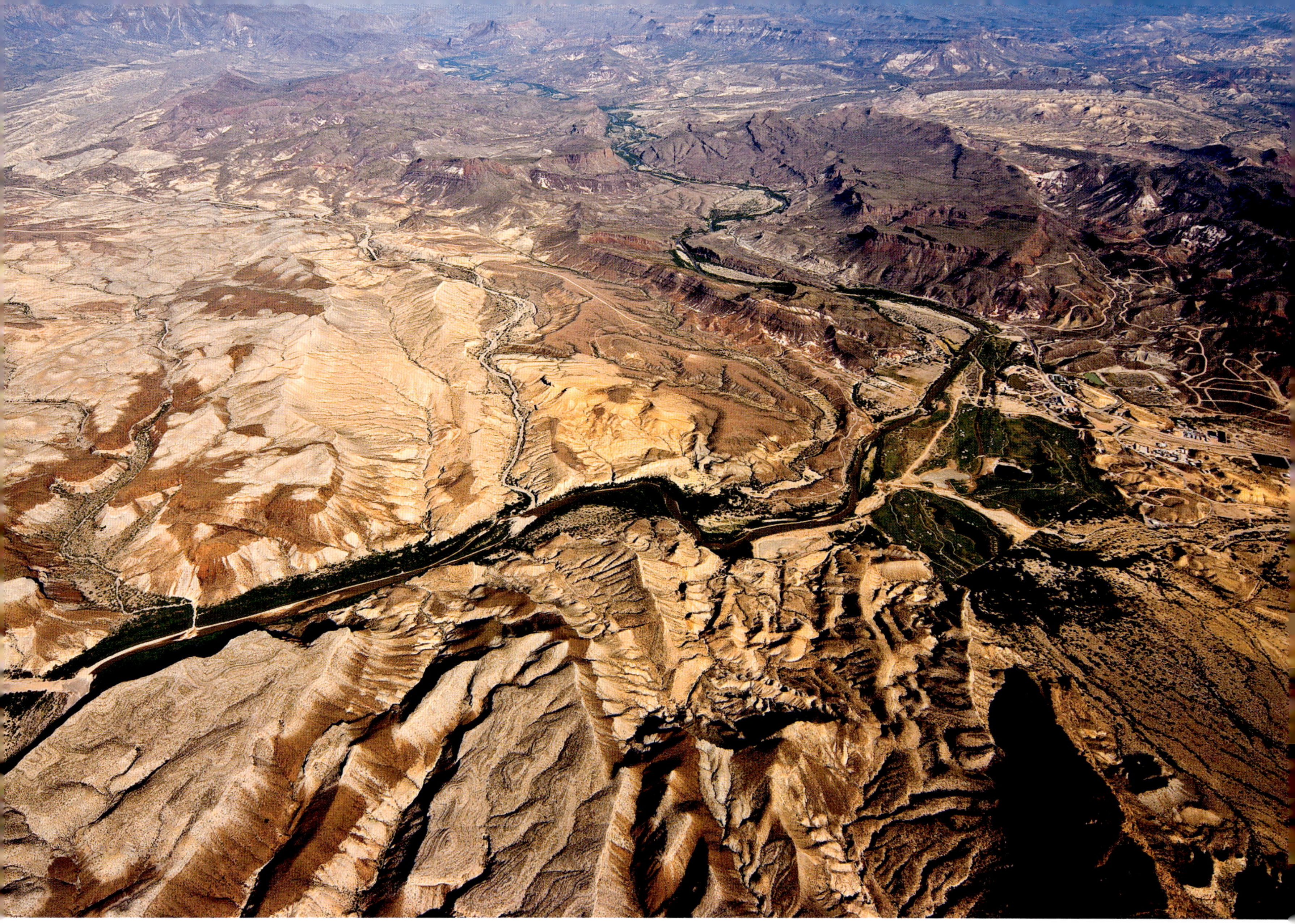

81. Resort Golf Course

Lajitas

Just outside the boundaries of Big Bend National Park lies the town of Lajitas (right), located on a bluff overlooking the Rio Grande. The Spanish word *lajitas* means “little flat rocks” and refers to the Boquillas limestone of the area that naturally paves the river bottom, making Lajitas the best river crossing around for scores of miles. Now a tourist destination with a 27,000-acre resort, spa, and golf course, Lajitas featured prominently in history as the site where the Great Comanche War Trail crossed the Rio Grande. It is also the place where Gen. John J. “Black Jack” Pershing set up a US Cavalry post to monitor the key river crossing as part of his campaign against Pancho Villa starting in 1916. I opted for this view to show the irrigated oasis of the resort surrounded by the sun-parched outback.

VII. IN SEARCH OF GHOST TOWNS

Flying Inland to the Lone Star and Mariposa Quicksilver Mines

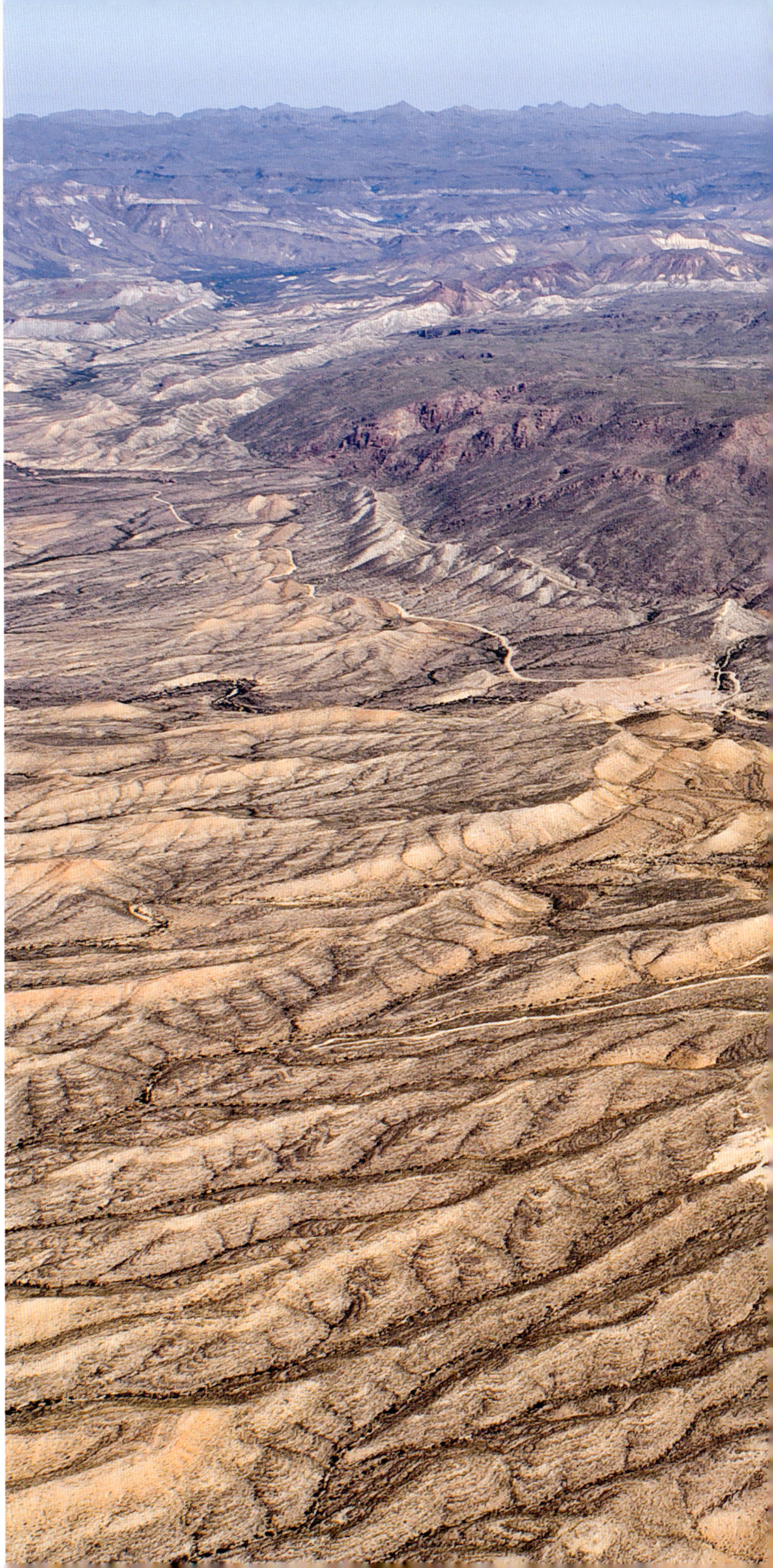

82. Lone Star Mine Road

Northeast of Lajitas

To the right, a dirt road switchbacks up Tres Cuevas Mountain to the area of the Lone Star Mine. To the left of center is the diminutive Amarilla Mountain, named after the Spanish word for yellow. Running through the center of the photograph is North Lone Star Mine Road, heading toward Contrabando Creek and Big Bend Ranch State Park in the distance. I was enamored by the visual motion in the sea of light-colored hills (left) that looked almost fluid, like waves lapping at the shore of dark rock (right).

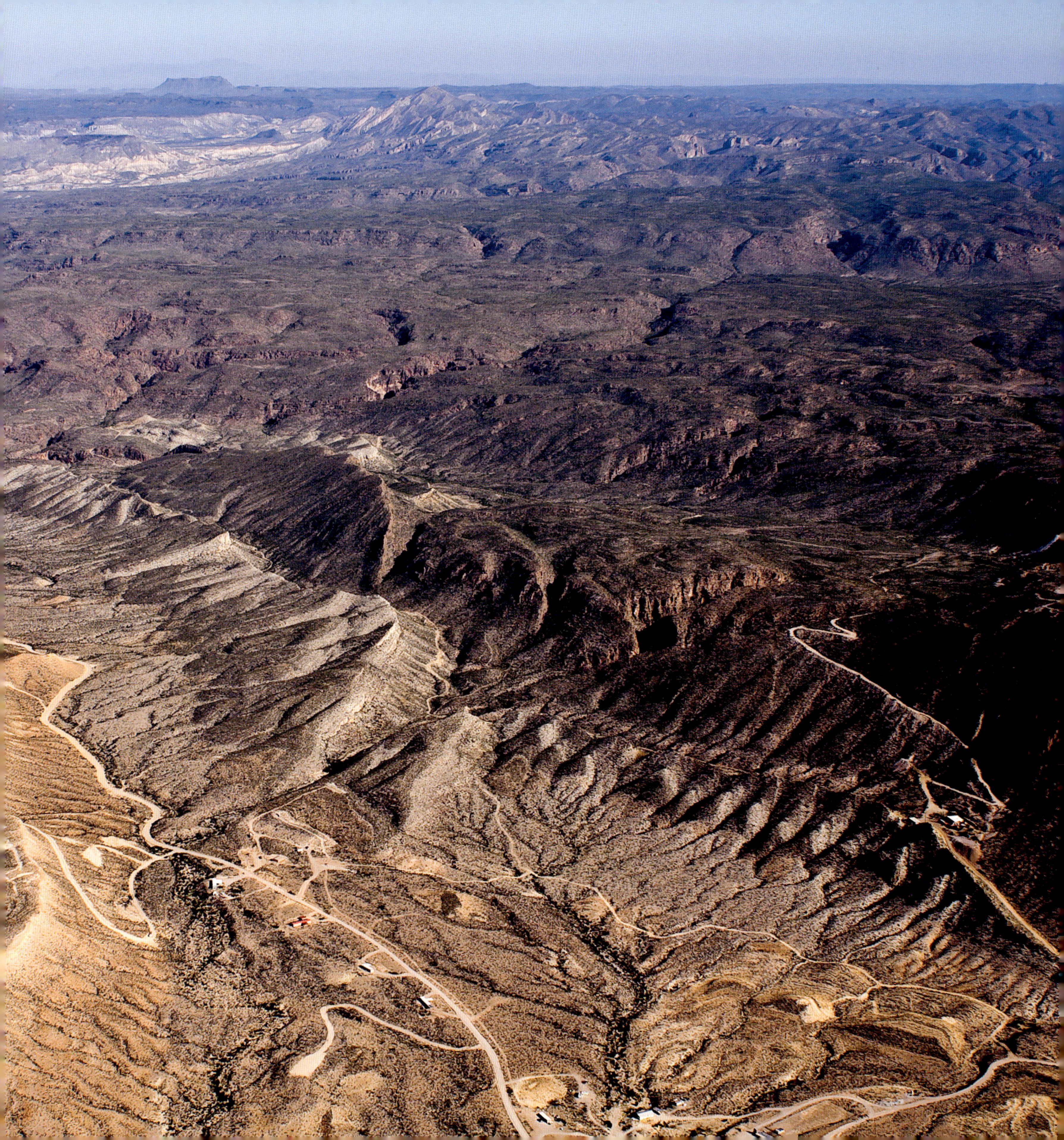

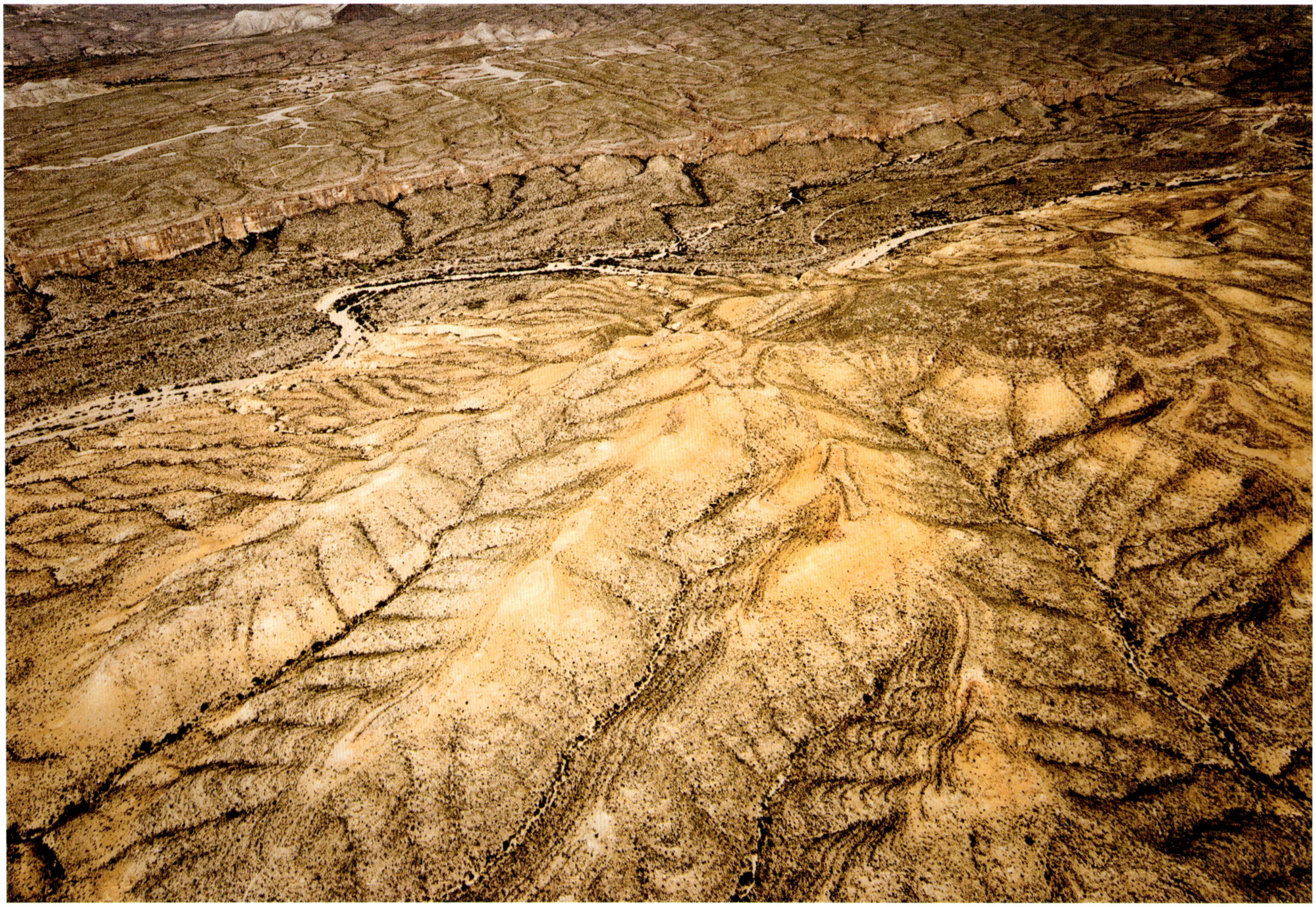

83. Terlingua Hills

West of Terlingua

Looking west across the Long Draw just north of Thirty-Eight Hill, one sees the area of Villa de la Mina in the distance. Villa de la Mina was the site of one of two famous chili cook-offs in the Terlingua area along Texas FM 170. The gently undulating hills in the Terlingua area are dotted with defunct mine shafts where cinnabar ore was extracted from the earth.

84. Villa de la Mina

Near Terlingua mining district

The apparent red tailings of an abandoned cinnabar mine (left) can still be seen at Villa de la Mina. Toxic mercury compounds resulting from cinnabar mining and mercury processing were typically left on-site. It is estimated that the Terlingua district has greater than two million cubic meters of mine waste, with mercury levels varying widely from 4.1 to 19,000 micrograms per gram. There are environmental concerns about the mercury compounds entering the watershed and being transported downstream toward Big Bend National Park. In previous times Native Americans used cinnabar as a red pigment for making pictographs and face paint.

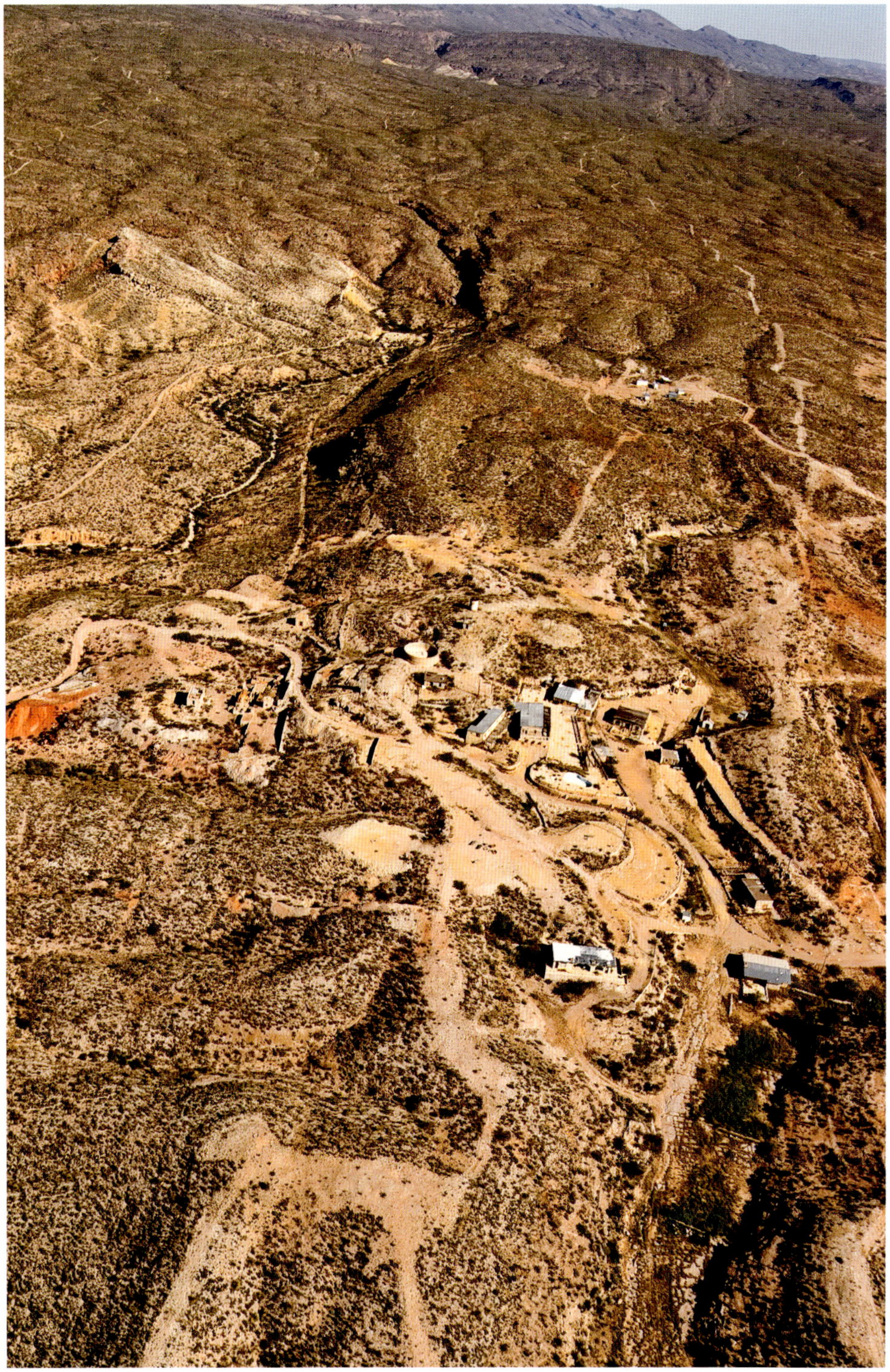

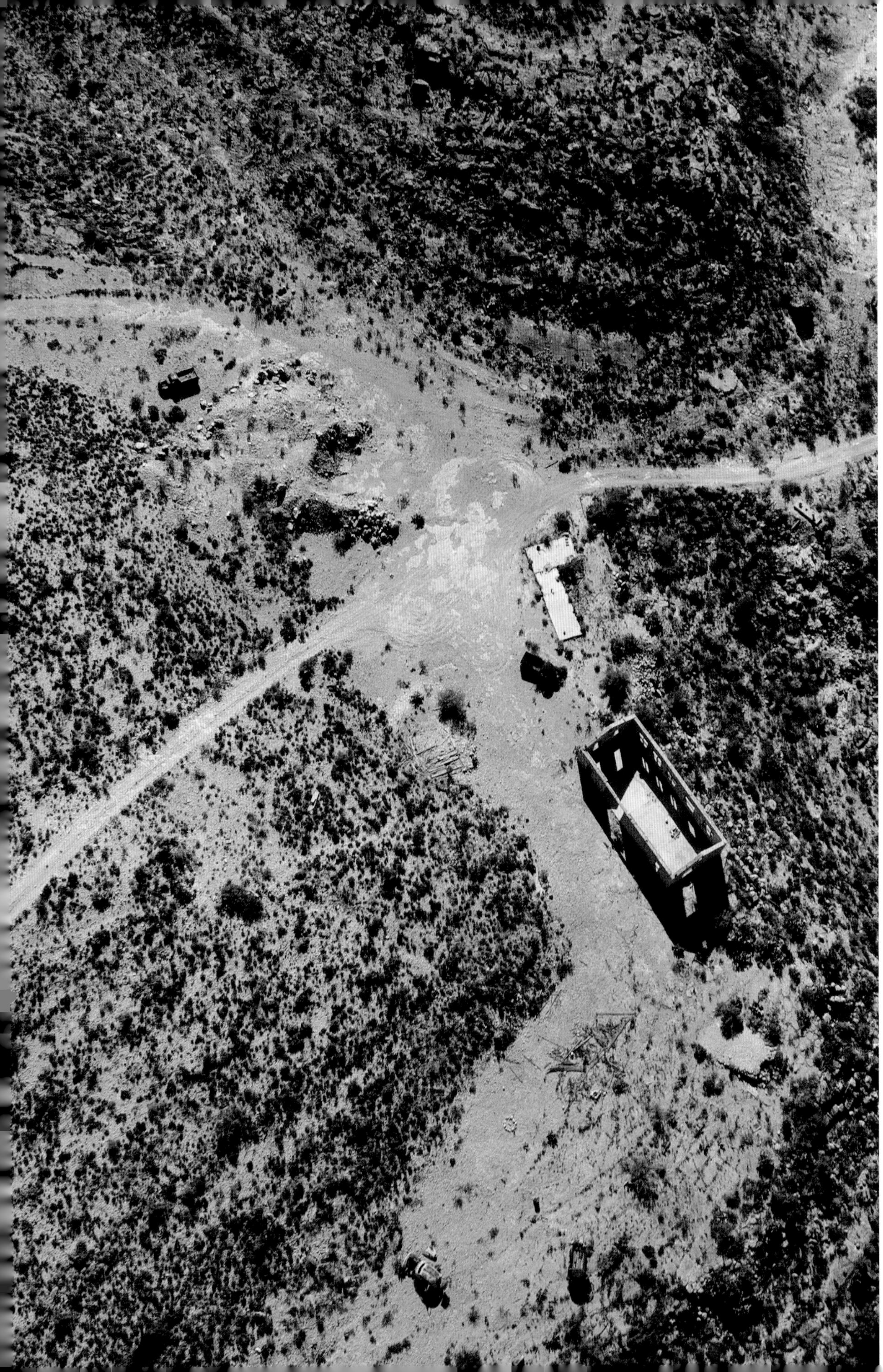

85. Ghost Town

South of Lowes Tinaja

The ruins of a building and rusting vehicles litter this tiny ghost town just south of Black Mesa in the Lone Star Mine district. After the decline of quicksilver mining in an area, buildings were frequently abandoned. The artifacts of the people who worked the mines, and often died at a young age, dot the hillsides along with their grave markers.

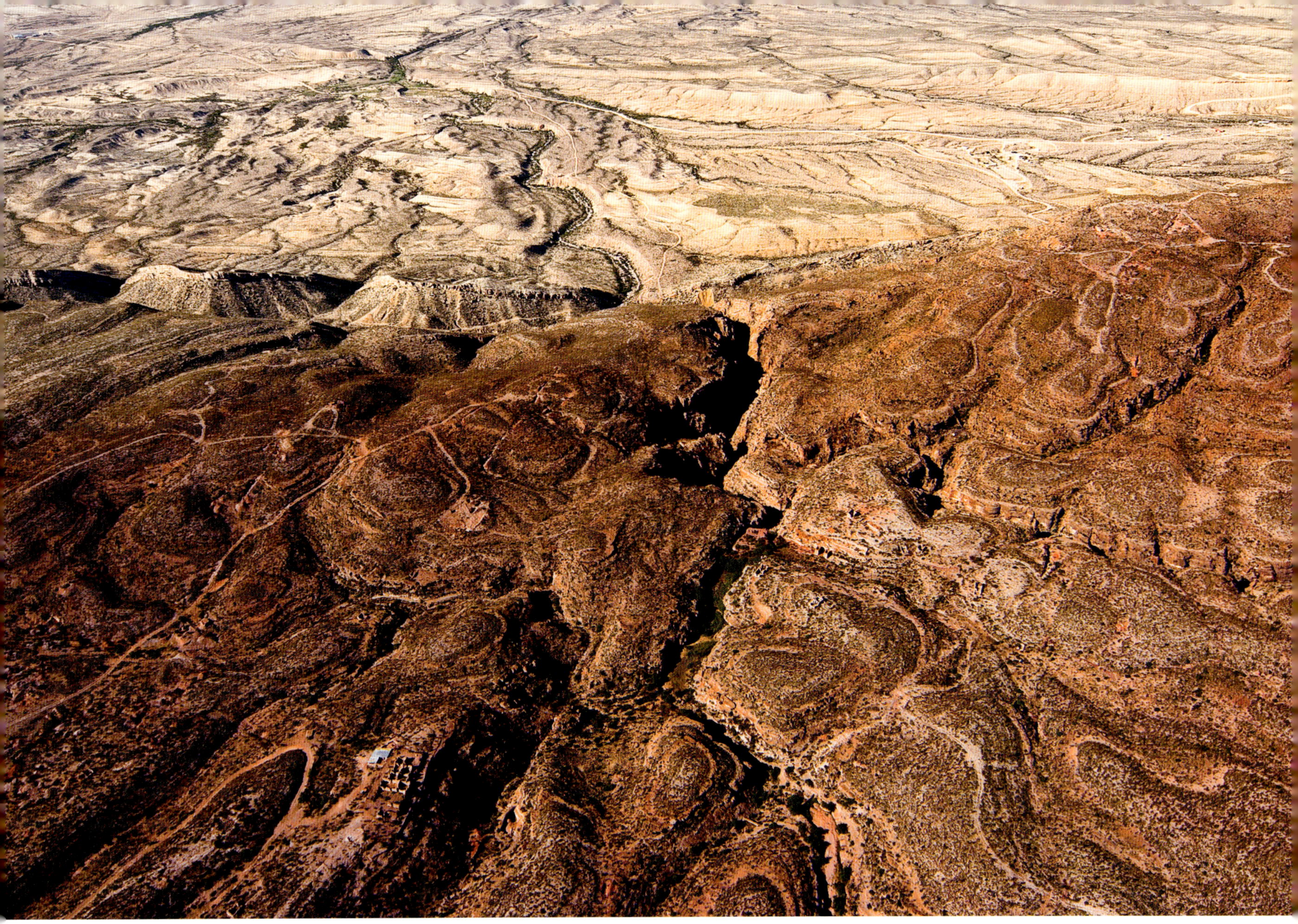

86. Quicksilver Mines

North of Lajitas

Mariposa Mine (left) and Lone Star Mine (right) were large quicksilver mines. The prominence to the left is California Hill, nearly circled by mine roads. Comanche Creek (center) cuts a canyon through the Sierra de Cal as it heads toward Lajitas, where it meets the Rio Grande. The mountaintop in this area is pitted with mine shafts and adits, and ruins are abundant. The Lone Star and Mariposa mines have been reclaimed and their openings fitted with grates that allow bats to fly through while keeping curious human visitors out of the decaying underground workings that are subject to cave-ins and dangerous gases.

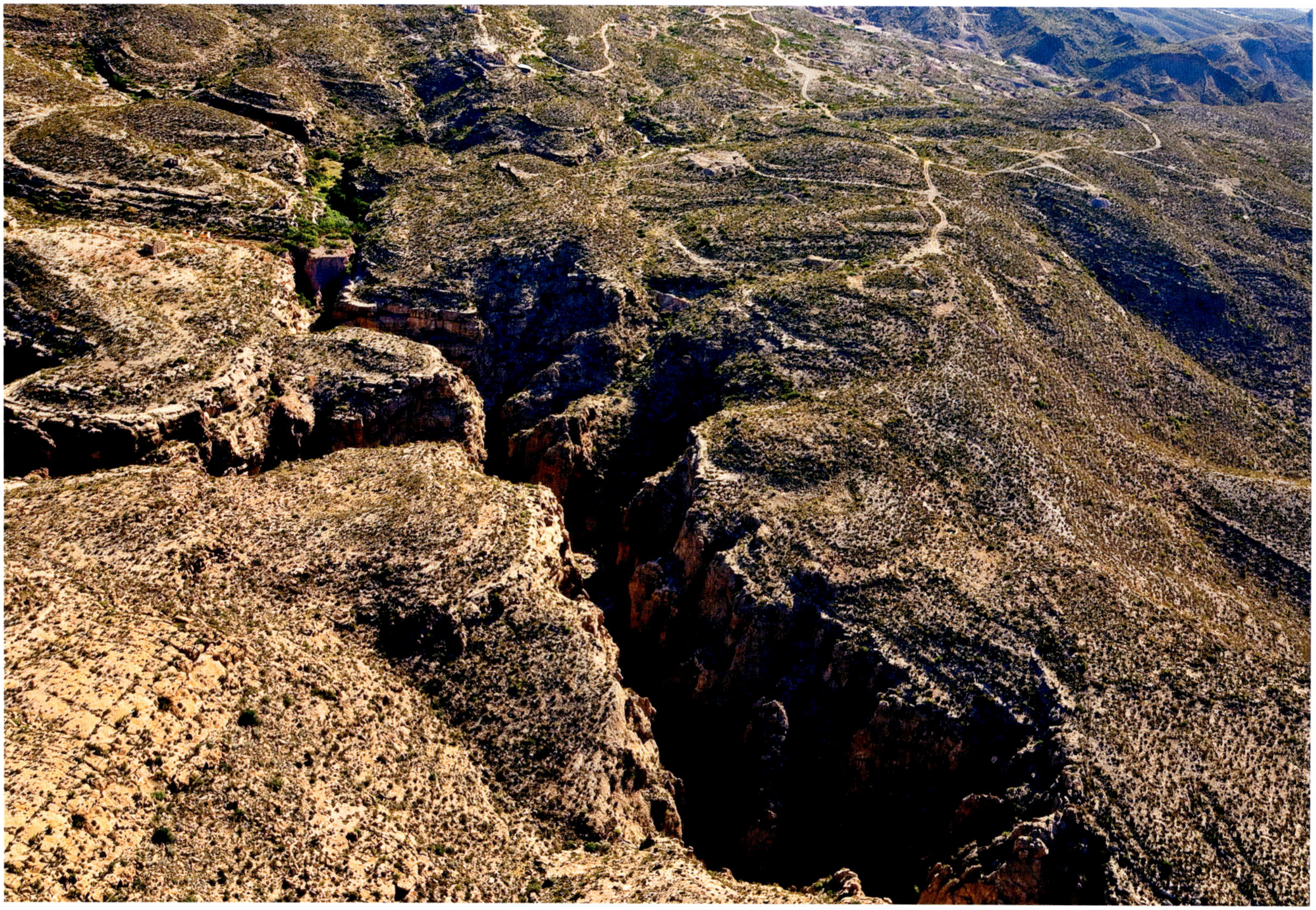

87. Canyon amid the Mines

Headwaters of Comanche Creek

Croesus Canyon is located in the upper reaches of Comanche Creek between the Lone Star and Mariposa mines. Here a spring creates an oasis of green and a miniature sanctuary for wildlife on the parched mountaintop of the Sierra de Cal. The availability of rare water makes the upper canyon a prime location for bird watching in a remote setting filled with history. As a photographer, I am intrigued by the way the light ricochets around the interior walls of the canyon, imparting a warm cinnabar-colored glow.

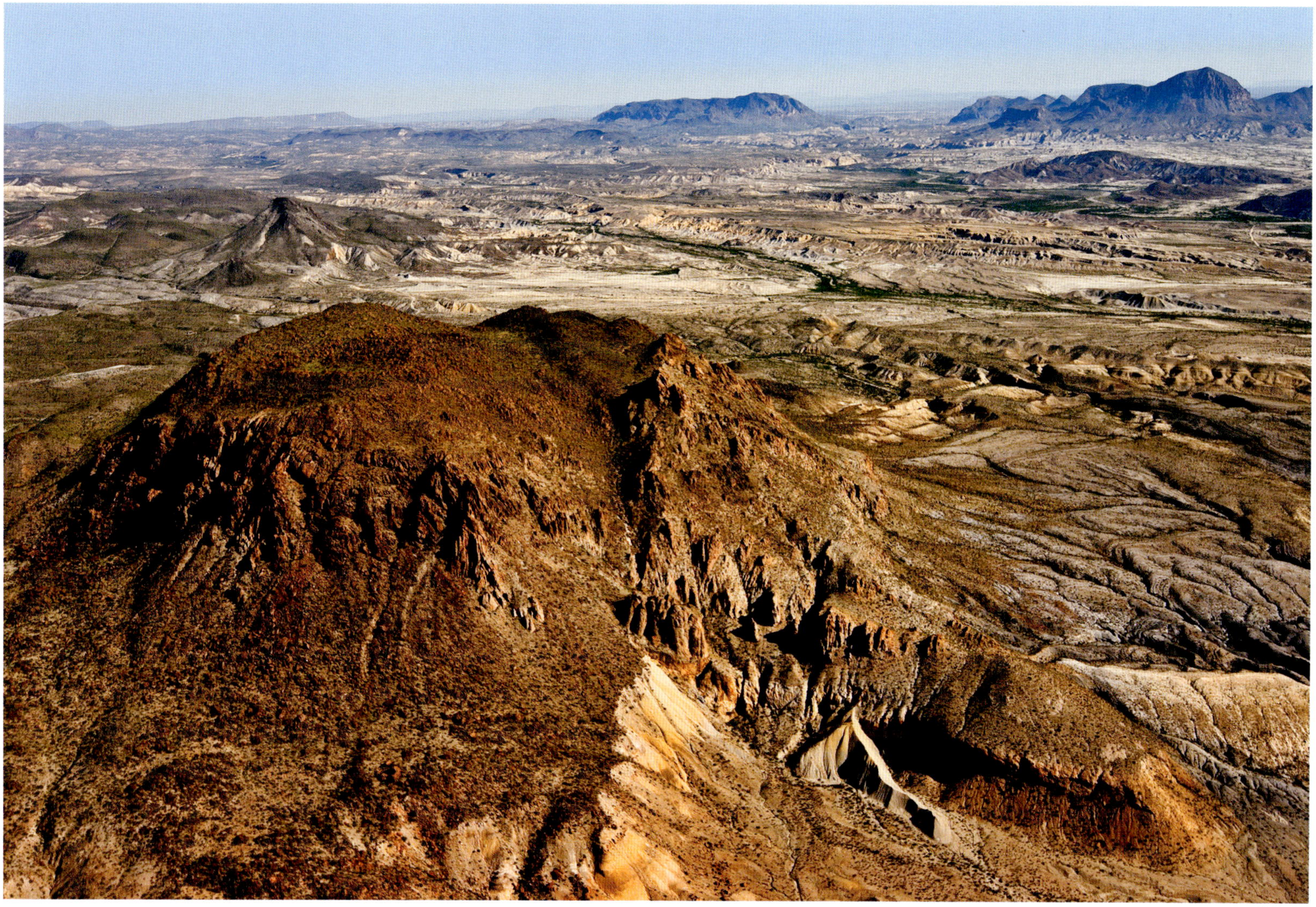

88. Sawmill Mountain

Three miles north of Terlingua

Sawmill Mountain (left), elevation 3,797 feet, rises sharply just west of Terlingua Creek as it flows south to the Rio Grande. Sawmill Mountain is composed mostly of intrusive tertiary igneous rock and is shaped by a normal fault that runs through the mountain. To the right is Texas Highway 118, which connects Terlingua and Study Butte with Alpine to the north. Sawmill Mountain is a key component in the aquifer system that flows through igneous rock and sedimentary limestone below the surface in this area. In the distance (right to left) are Hen Egg Mountain and Agua Fria Mountain. The humps of the mountains echo each other visually, creating visual rhythm while drawing our gaze into the seemingly interminable distance of the desert.

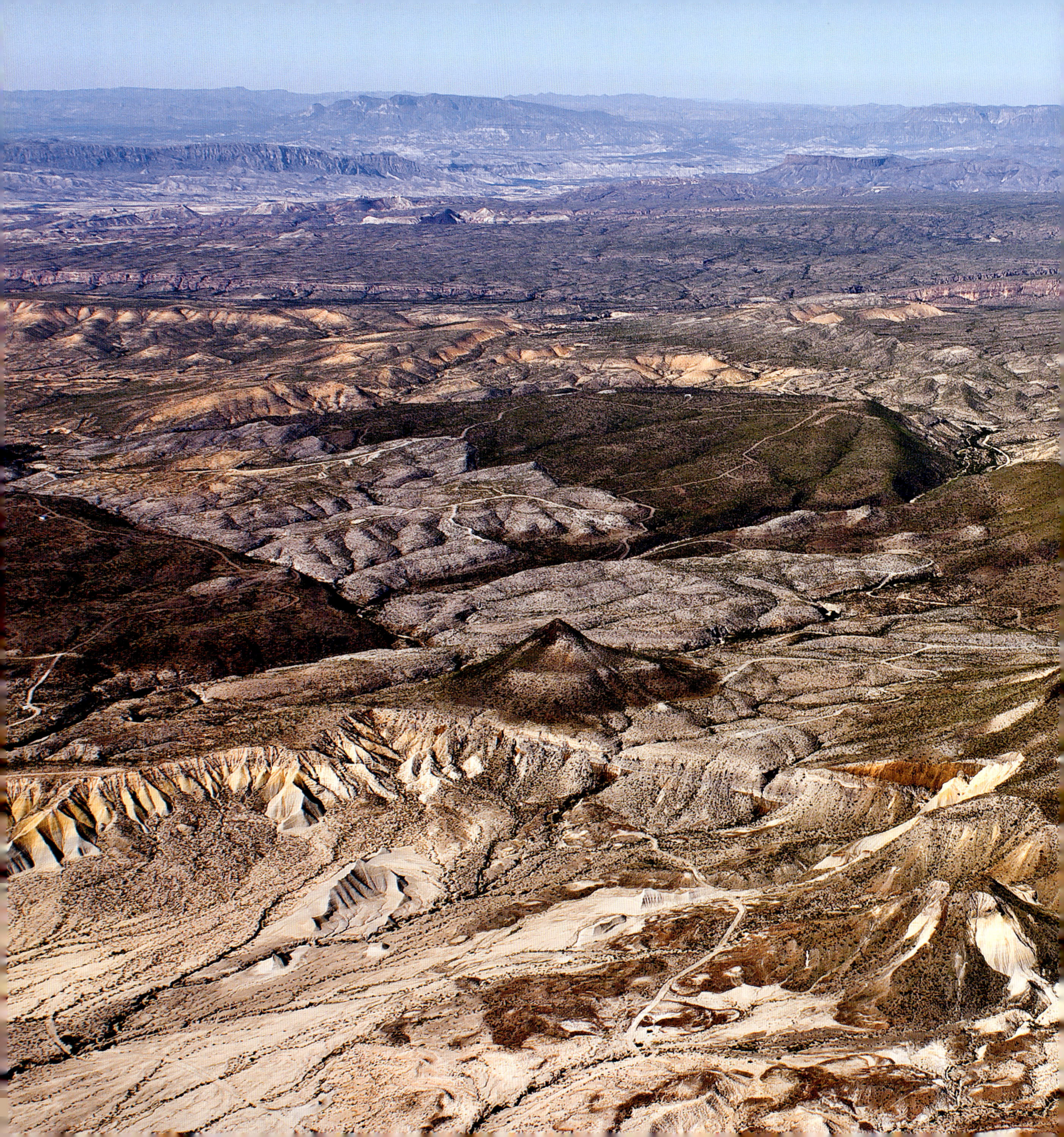

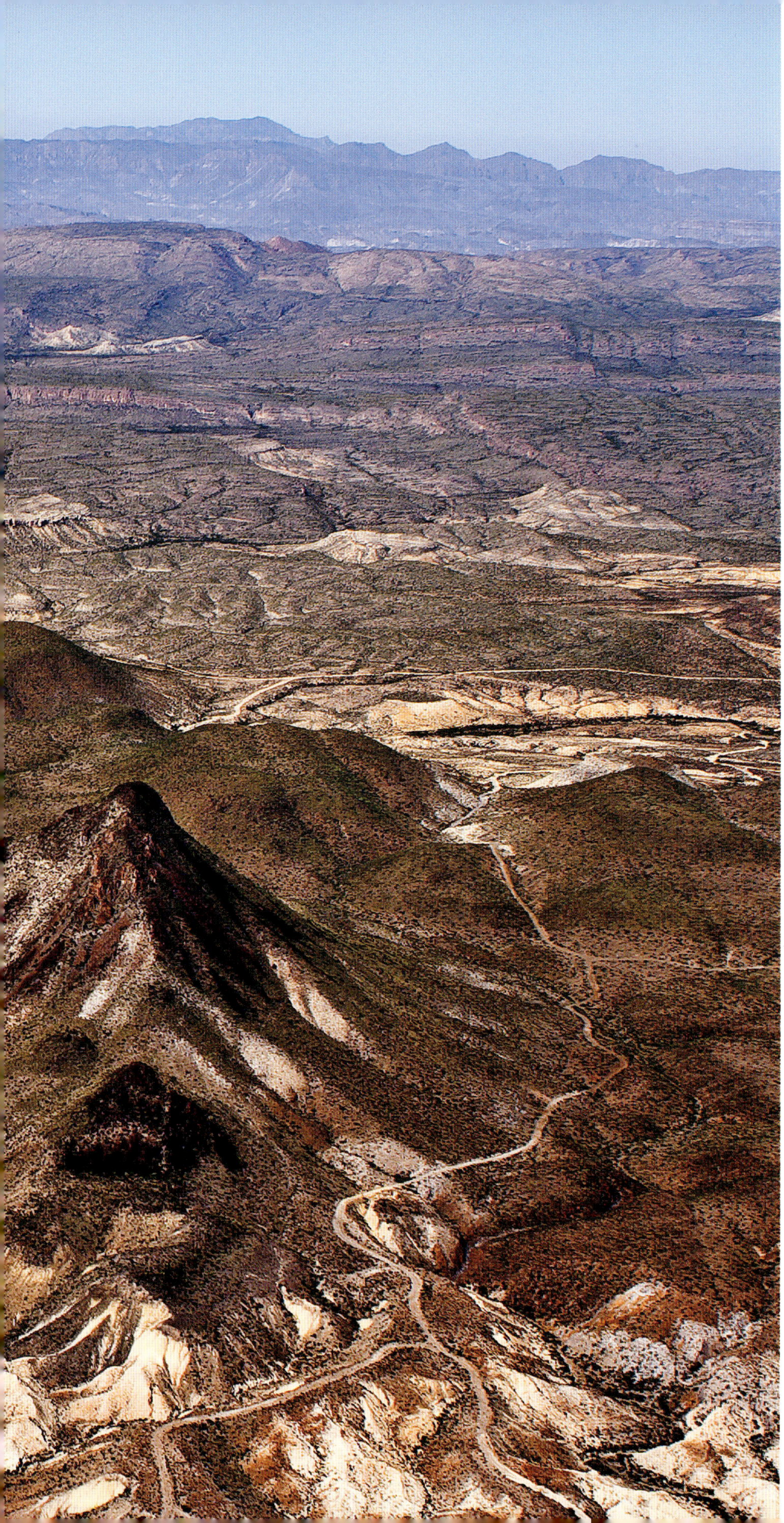

89. Pinks Peak

Two miles west of Terlingua Creek

Pinks Peak (lower right) is named for the Big Bend rancher Pink Taylor. In 1895 Taylor purchased a large number of cattle from the famous G4 Ranch when it disbanded. At its zenith, the G4 Ranch ran one of the biggest West Texas ranching operations, with 30,000 head of cattle, between here and the Chisos Mountains. The G4 Ranch took its name from the survey block where it was located and comprised 55,000 acres. In addition, the G4 leased adjacent land with water from the railroad, which controlled most water rights in its land grants.

VIII. INTO THE DEVIL'S PLAYGROUND

Flying North from Terlingua to Alamo de Cesario Creek

90. Strawberry-Colored Tuff

At the Presidio-Brewster county line

Colorful layers of volcaniclastic ash, known locally as strawberry and purple tuffs, paint the fantastic hills south of Bandera Mesa. This view is looking southwest toward Big Bend Ranch State Park, Alamos Spring (right), and Panther Canyon (upper left). I took this photograph to invite the viewer to share in this vista of enchanting colors and shapes delightfully punctuated by Alamo de Cesario Creek.

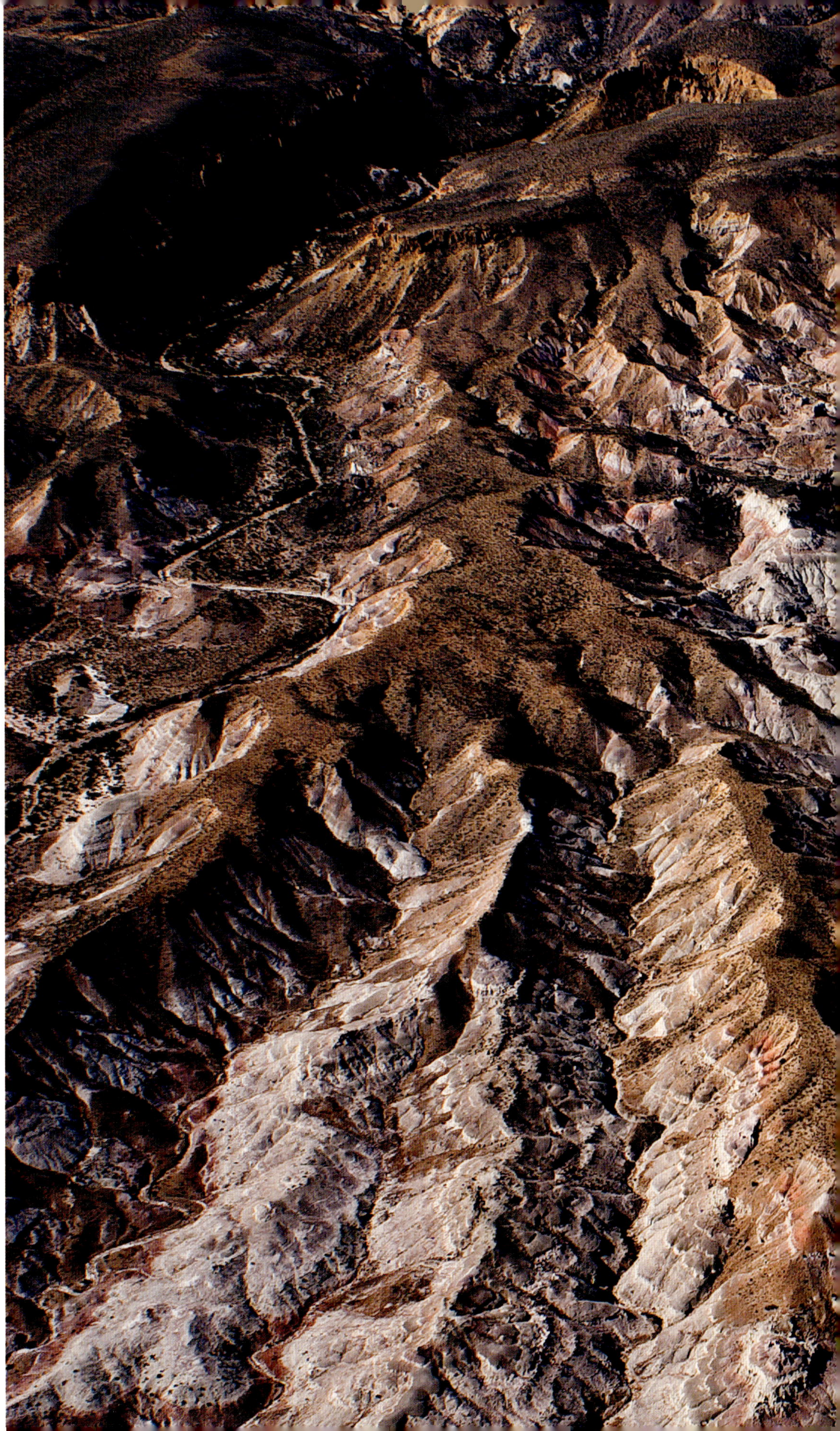

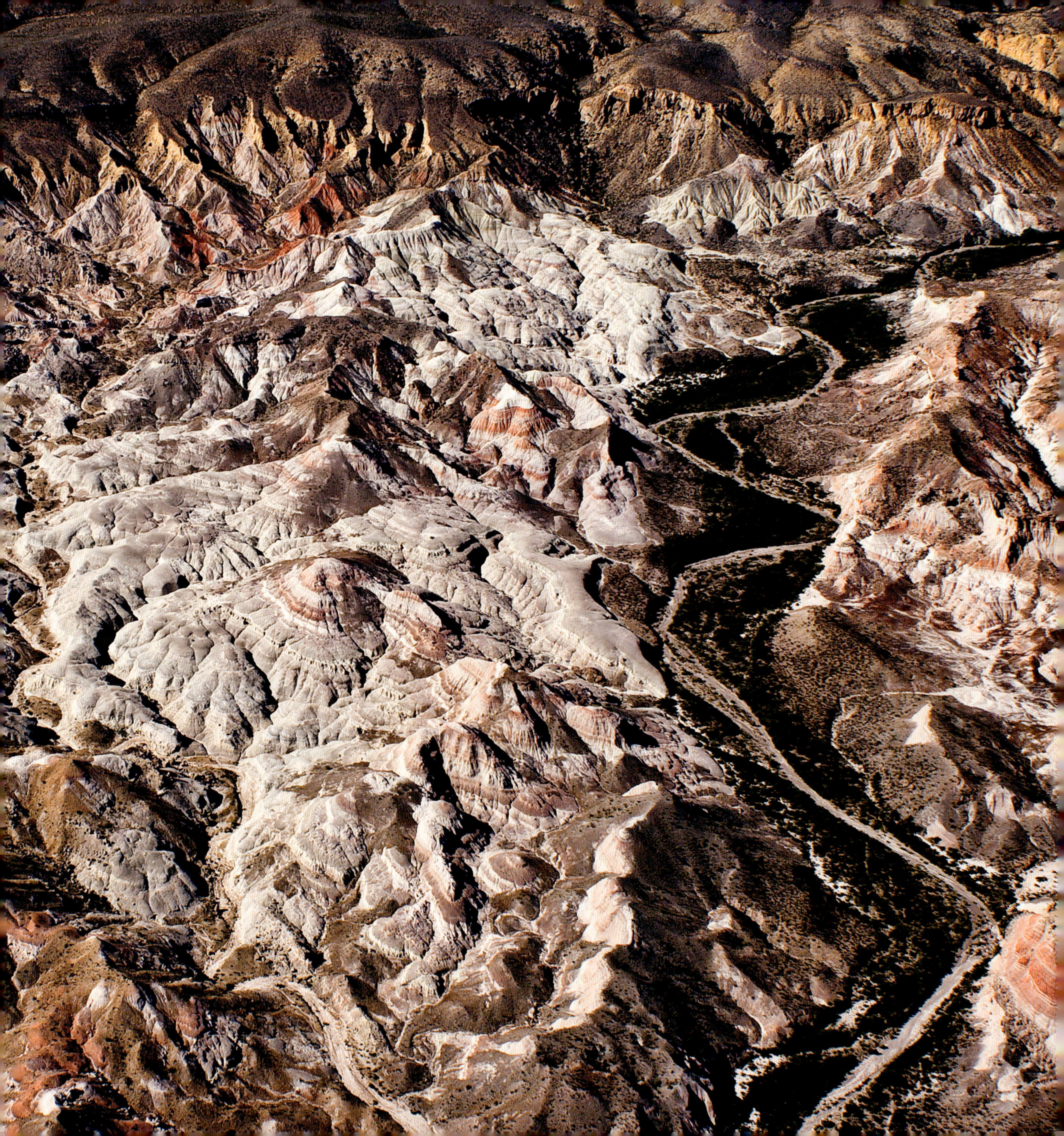

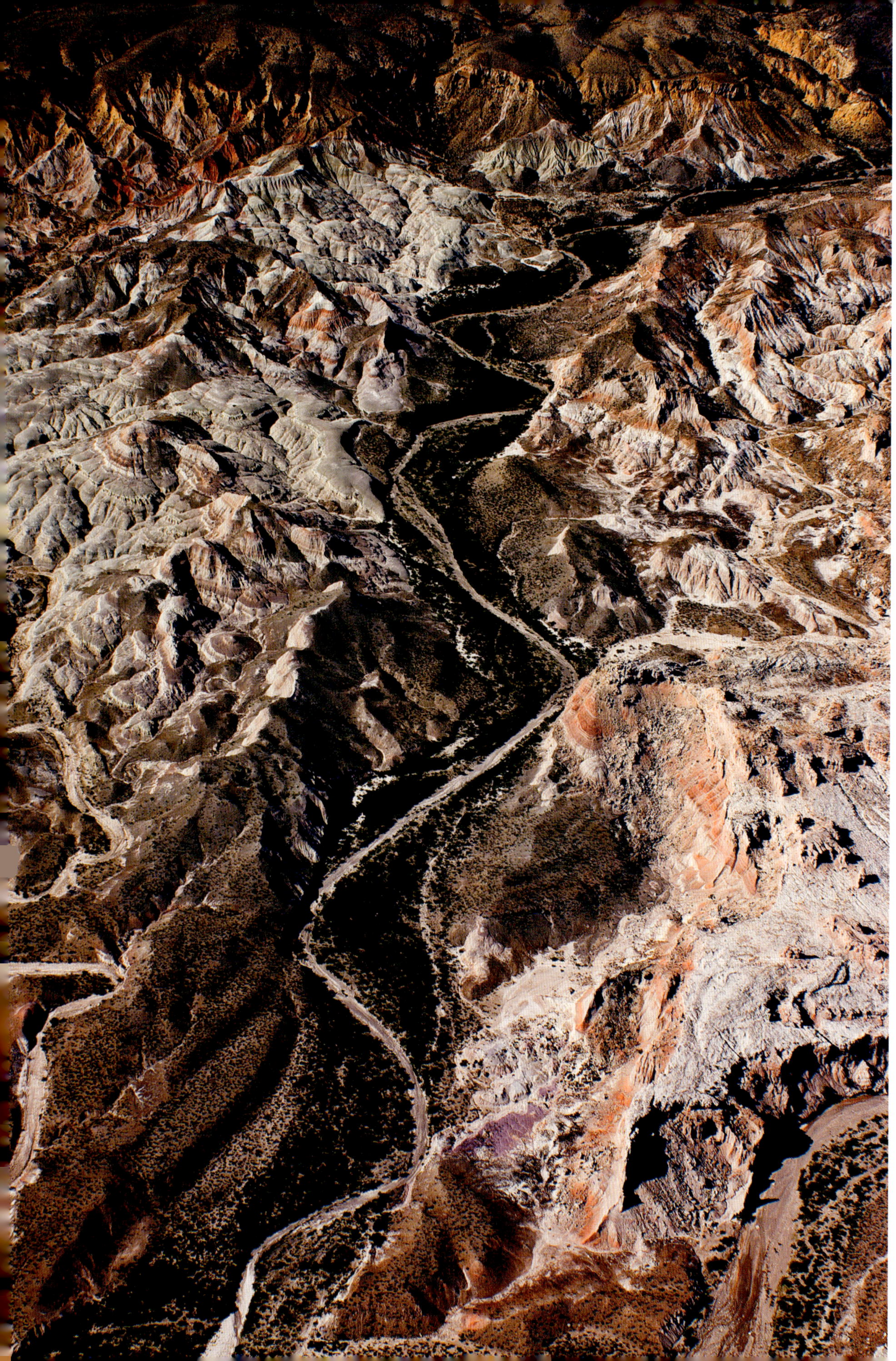

◀ 91. Zigzagging Creek

South of Bandera Mesa

The south fork of Alamo de Cesario Creek, near its confluence with the main stream, is lined with desert conifers and oak trees. The serenity and beauty of these hills contrast with their origin as incendiary ash falling from an erupting volcano ages ago. I included this close-up to emphasize the visual poetry formed by the interaction of water and land.

▶ 92. Canyon Detail

Alamo de Cesario Creek

At the edge of the Bandera Mesa (left) just west of Cottonwood Spring, Alamo de Cesario Creek dives into a serrated canyon where the watercourse strikes an unnamed 3,996-foot-high point and is forced to make a ninety-degree turn to the right. Erosion reveals layers of the saturated color of sand, mudstone, and tuff in the walls. After desert rainstorms, the runoff from Bandera Mesa turns the creek into a raging flash flood. The bedazzling texture of the hillsides drew me to this scene, along with the slalom-course line of the creek bed.

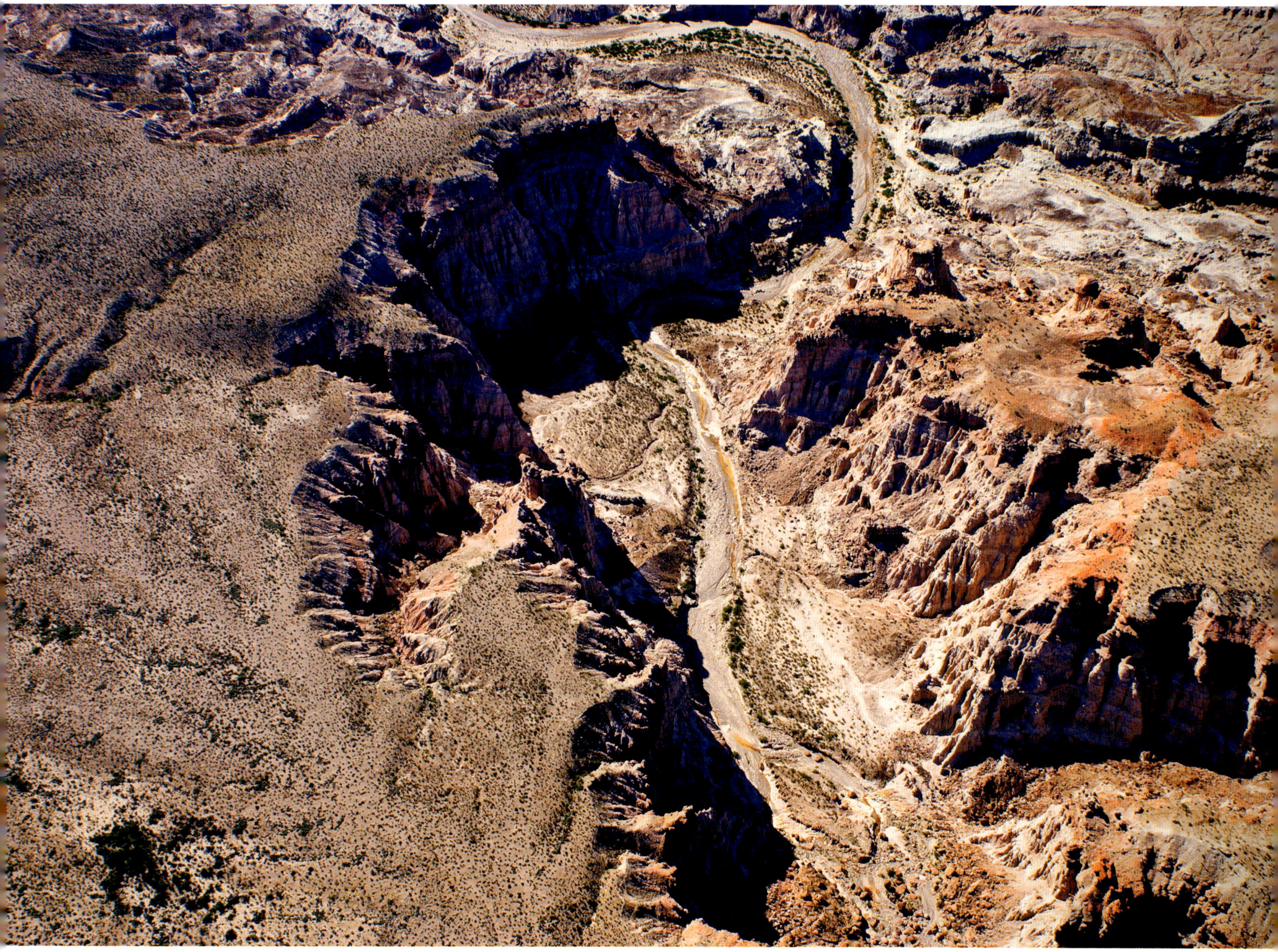

93. Badlands of the Devil's Playground

North of Alamo de Cesario Creek

This half-mile clump of earth reveals the eye-popping conglomeration of textures and geological forms at the edge of Bandera Mesa. Looking straight down on the land in the vertical aerial photograph reveals the dizzying color and assortment of forms created by the full spectrum of nature's forces. Due to its hellish terrain, the place is locally known as the Devil's Playground.

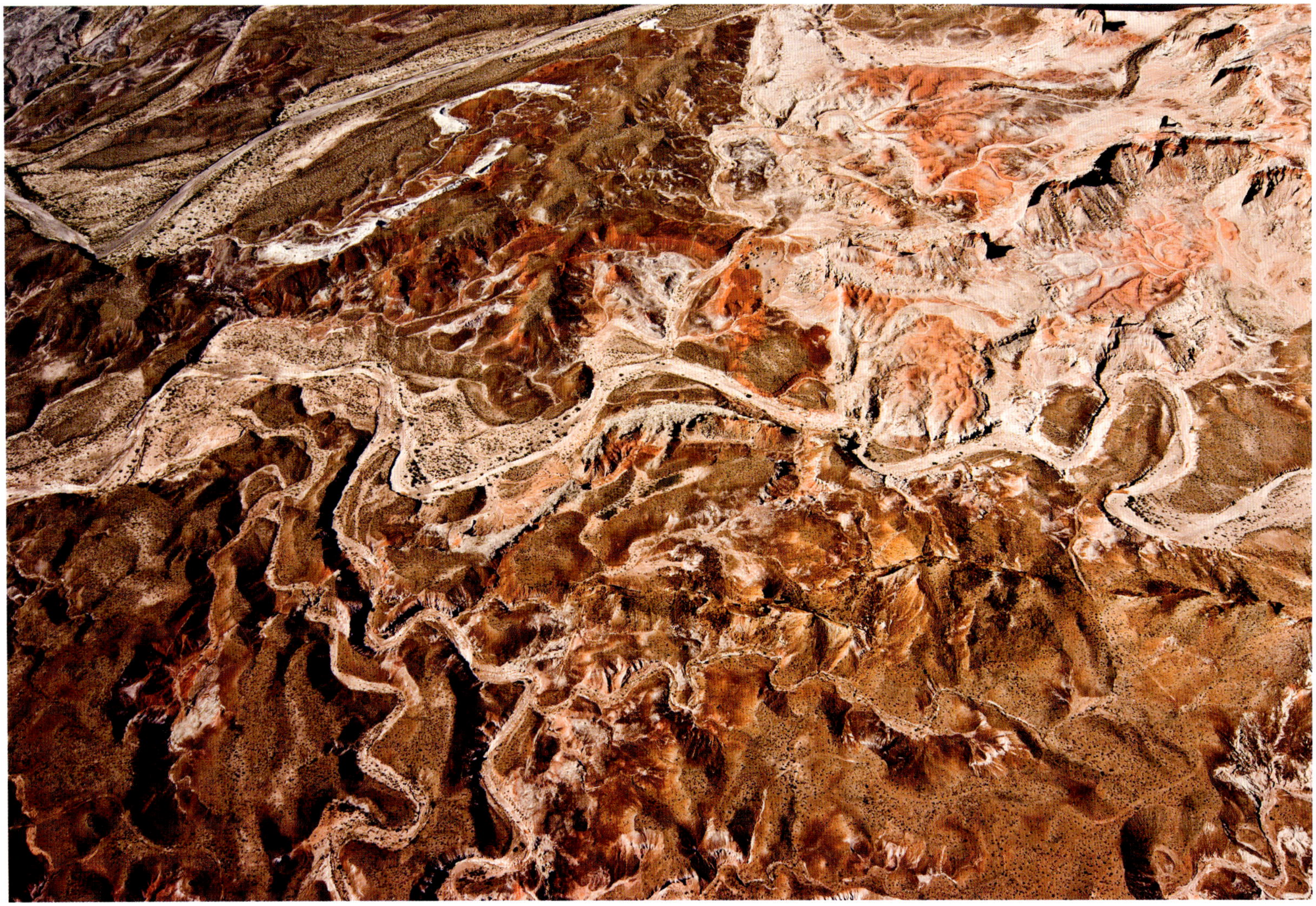

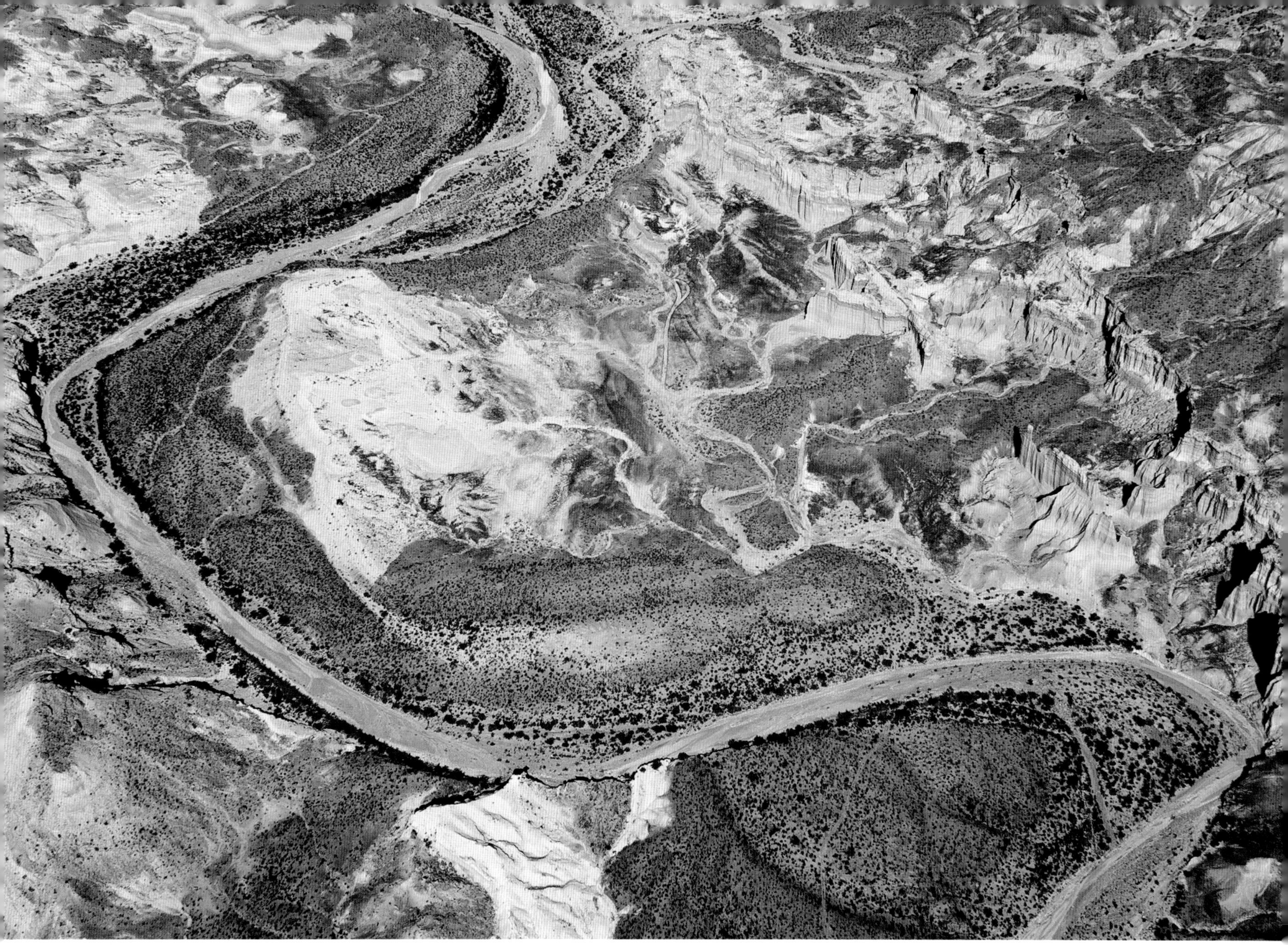

94. Creek Curves Detail

Alamo de Cesario Creek

The riverbed here follows the path of least resistance around numerous obstructions in the landscape, resulting in a graceful yet squarish curve across the desert dry wash. Facing an obstacle at every turn, the creek changes trajectory as it finds its way downstream toward Terlingua Creek, twelve miles away. The minerals in the land color the scene with rich reds, browns, contrasting white, and the tan alluvium of the draw.

95. Cesaria's Cottonwood

Just west of Fizzle Flat

The folk story about how Alamo de Cesario got its name reads like a romance novel. Once upon a time in Mexico there was a beautiful girl named Cesaria. Her homeland was raided by Mescalero Apaches who took her captive as they rode north back across the Rio Grande with their plunder. At Agua Fria they camped by the cool water of the spring and tied the maiden to a cottonwood tree, or *alamo* in Spanish. Her loved ones attacked the Indians, drove them off, and rescued her. Ever since that day, the area has been known as Cesaria's Cottonwood and on USGS topographic maps the stream to the north still bears her namesake, Alamo de Cesario. I took this photograph, beguiled by the rich tapestry of the landscape and the story of the place.

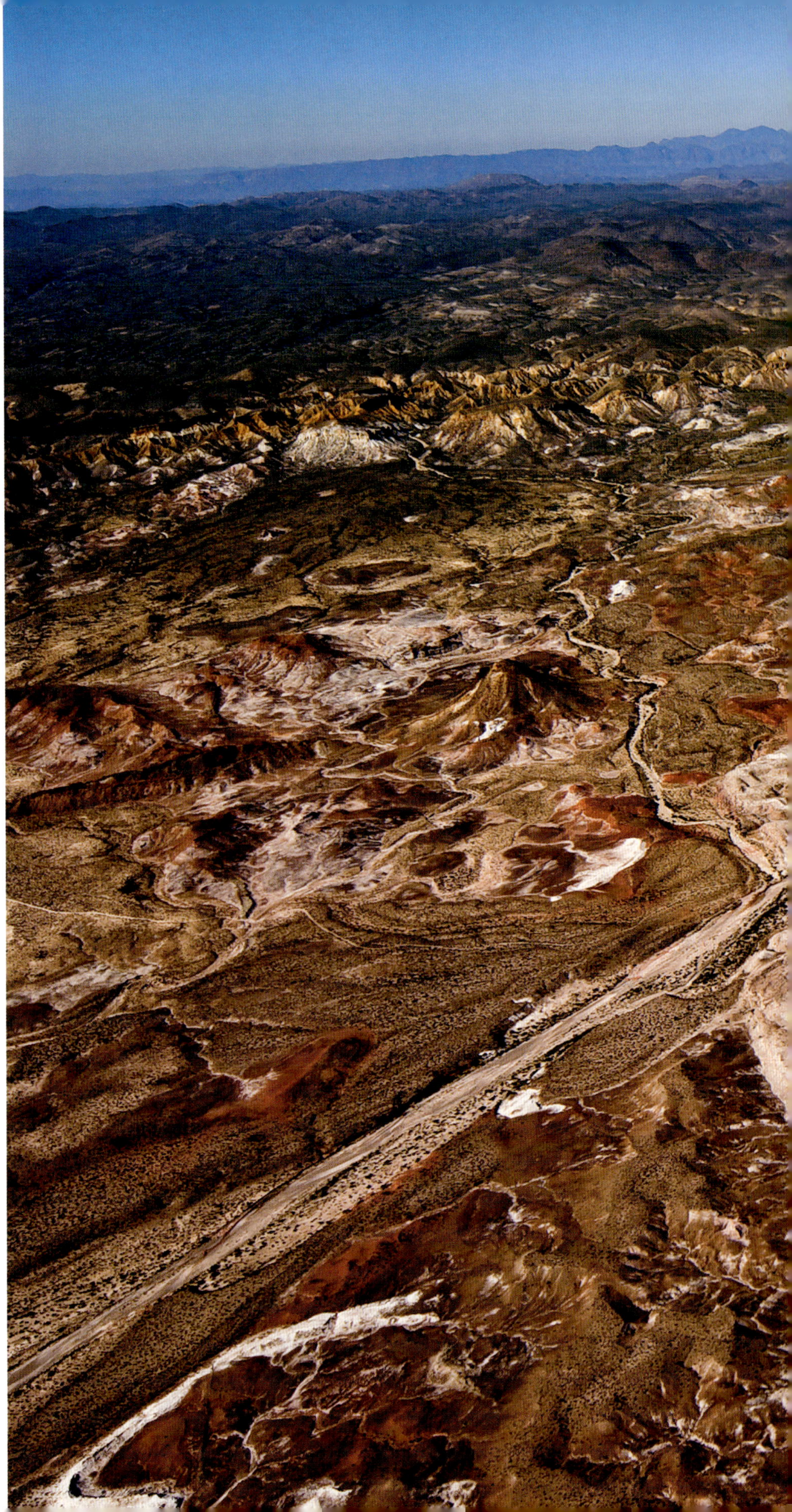

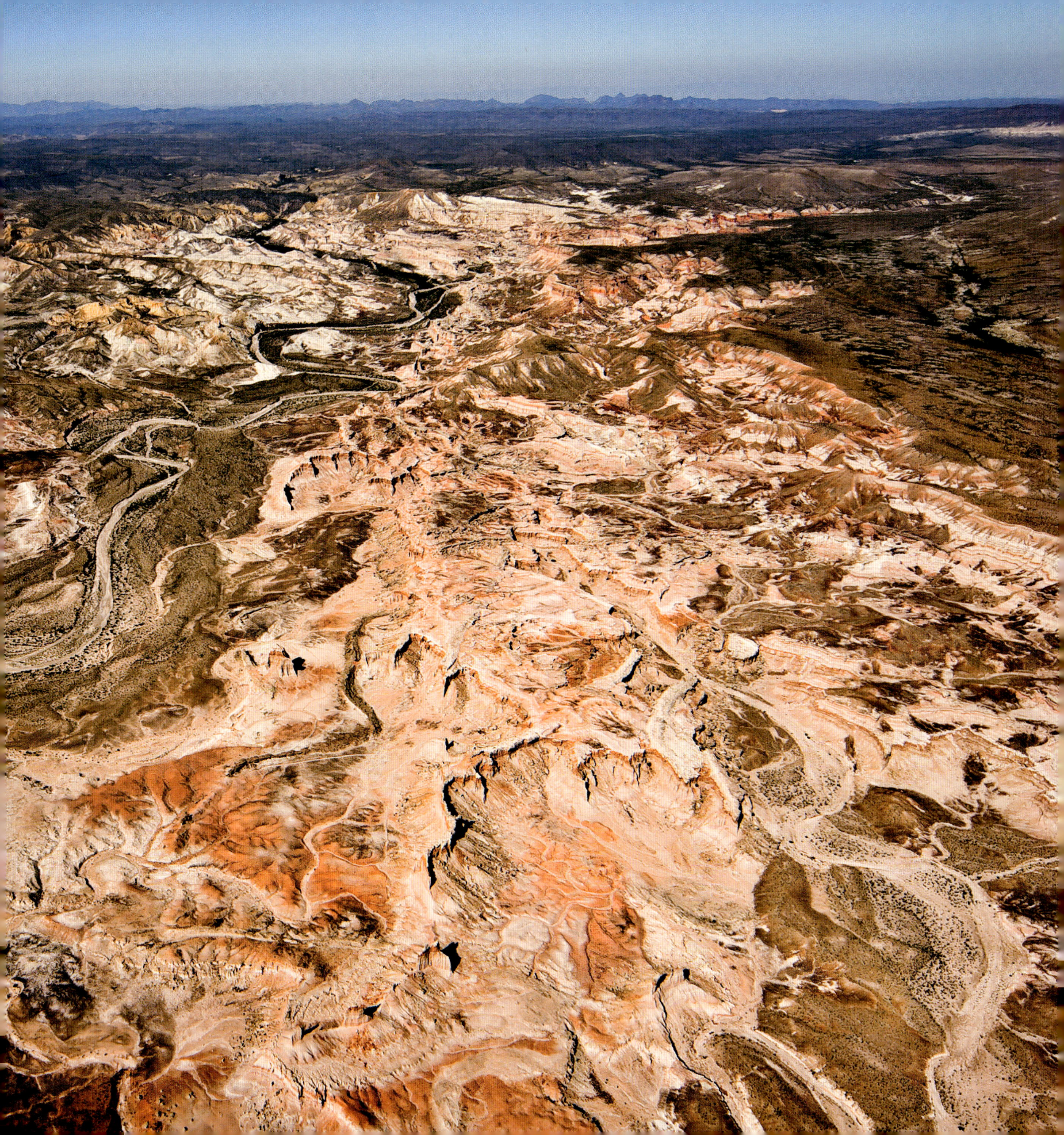

IX. OVER THE LAND OF PLENTY

Orbiting Green Valley and Paradise Valley

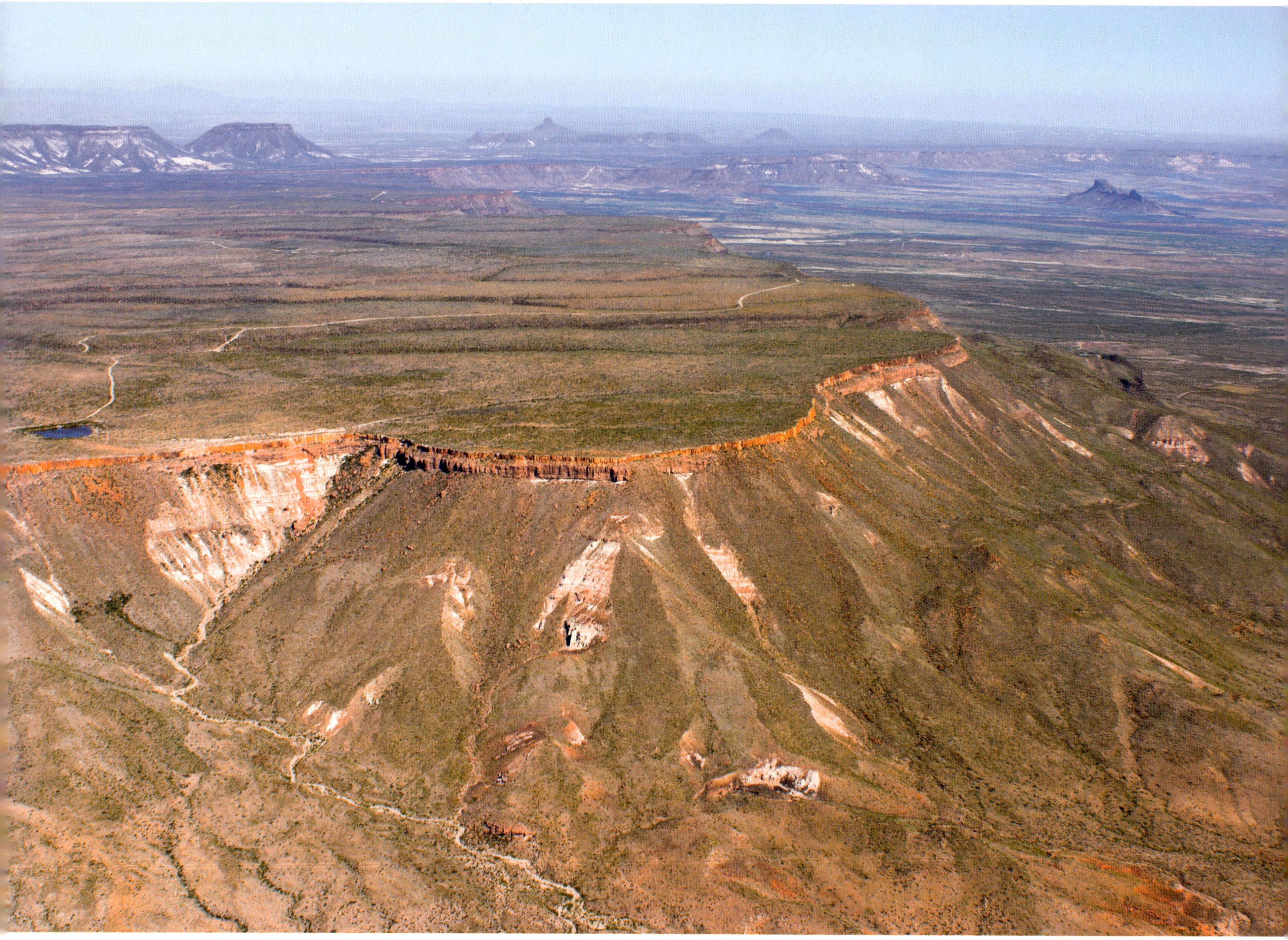

96. Bandera Mesa

About fifteen miles south of Plata

The plateau (left) is the broad flat top of Bandera Mesa, which measures about four miles wide and seven miles long. A stock tank (left) provides water for livestock and deer on the mesa. In the distance (right) are the relatively water-rich ranchland of Green Valley and the distinctive shape of Needle Peak. I relish the dreamy color of this lyrical pastoral scene.

97. Needle Peak

Green Valley

Needle Peak (center), altitude 4,265 feet, rises from Green Valley near Rancho del Martillo. The peak is a syenite dike formed by magma injected into a crack in Duff tuff. The rock pinnacles are approximately 80 percent feldspar. Other mountains in the distance (left to right) are San Jacinto Mountain and Cerro Boludo. To the right is another intrusion simply called the Dike.

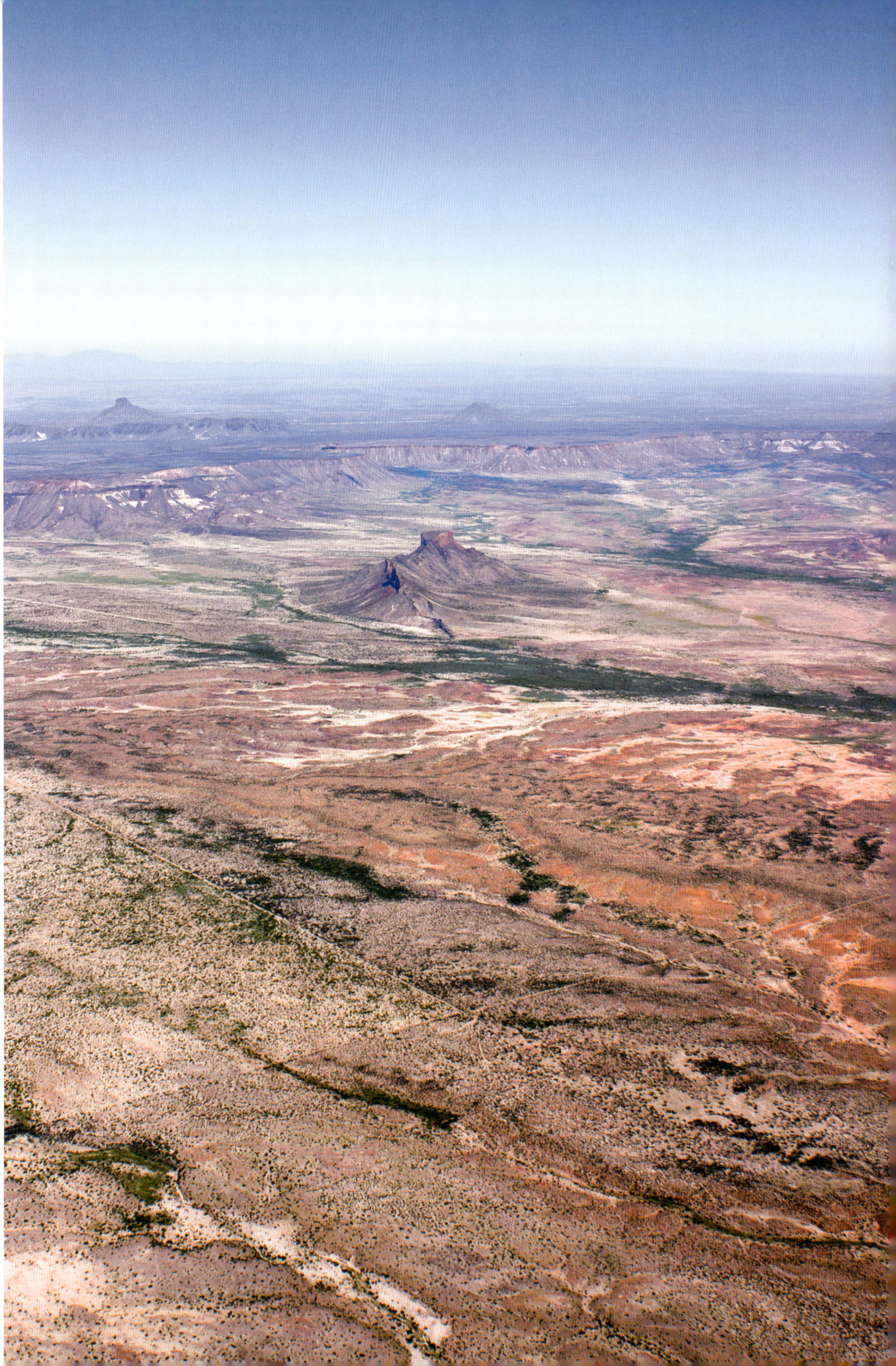

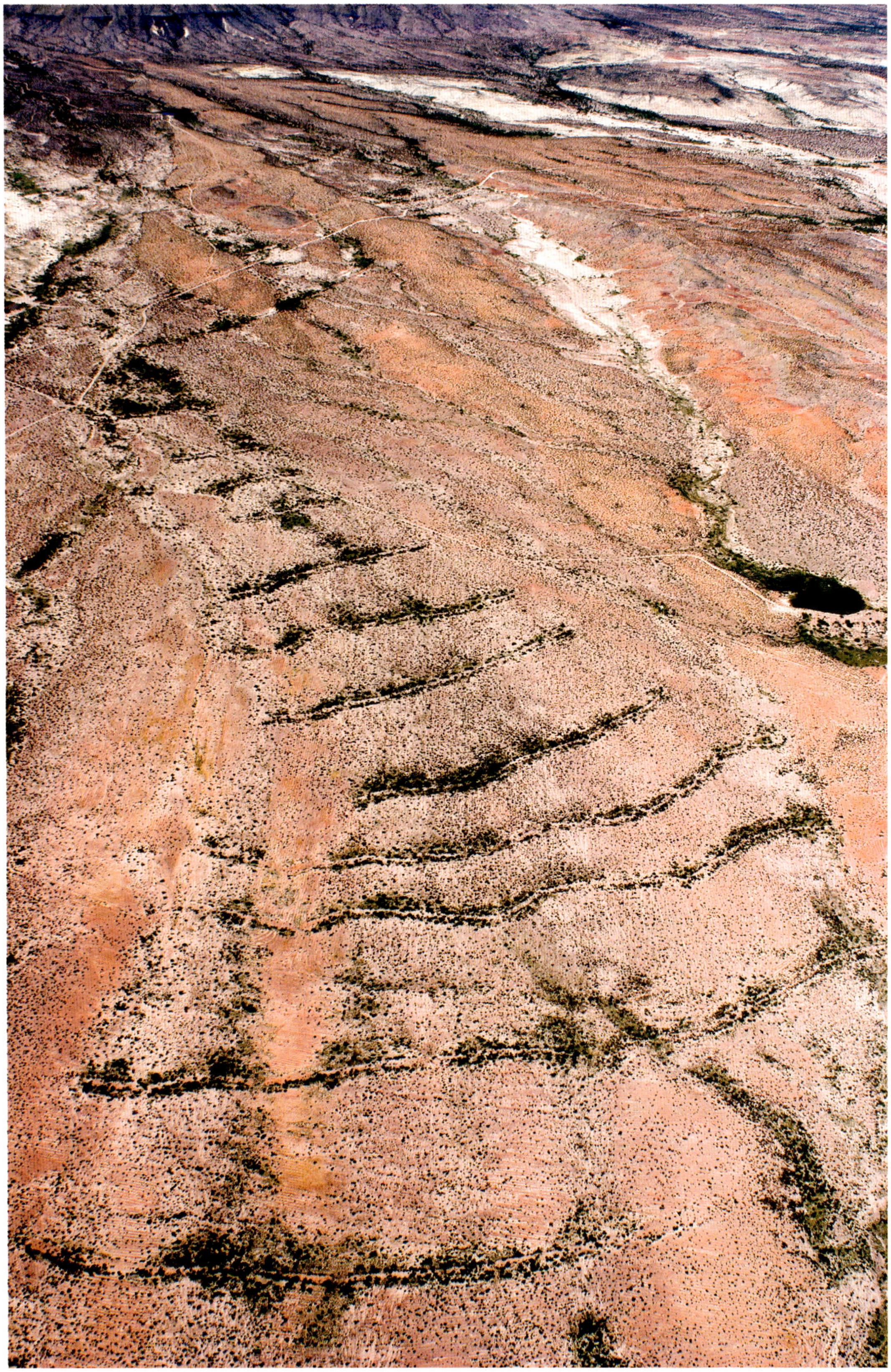

◀ **98. Green-Terraced Valley Floor**

Two miles northeast of Bandera Mesa

Terraces banded with vegetation and a stock tank ringed with green slow and catch the runoff from Bandera Mesa in Green Valley. The playful rippled pattern of green lines and the dot of water appearing in the desert basin add a whimsical note of life to the dry open range.

▶ **99. McKinney Mountain**

Thirteen miles west of Texas Highway 118

McKinney Mountain (center), 5,006 feet, and Turney Peak (right), 4,854 feet, dominate this view looking north-northeast toward Paradise Valley. Clegg Fowlkes operates Hammond Ranch in this area around McKinney Mountain. Beyond that are Paradise Draw and the upper reaches of Terlingua Creek. Distant peaks just to the right of center (top) include Goat Mountain, Cathedral Mountain, and Cienega Mountain. I gleaned this image from the bunch of pictures for the pure drama and monumental presence of this monolithic mountain rising from the valley.

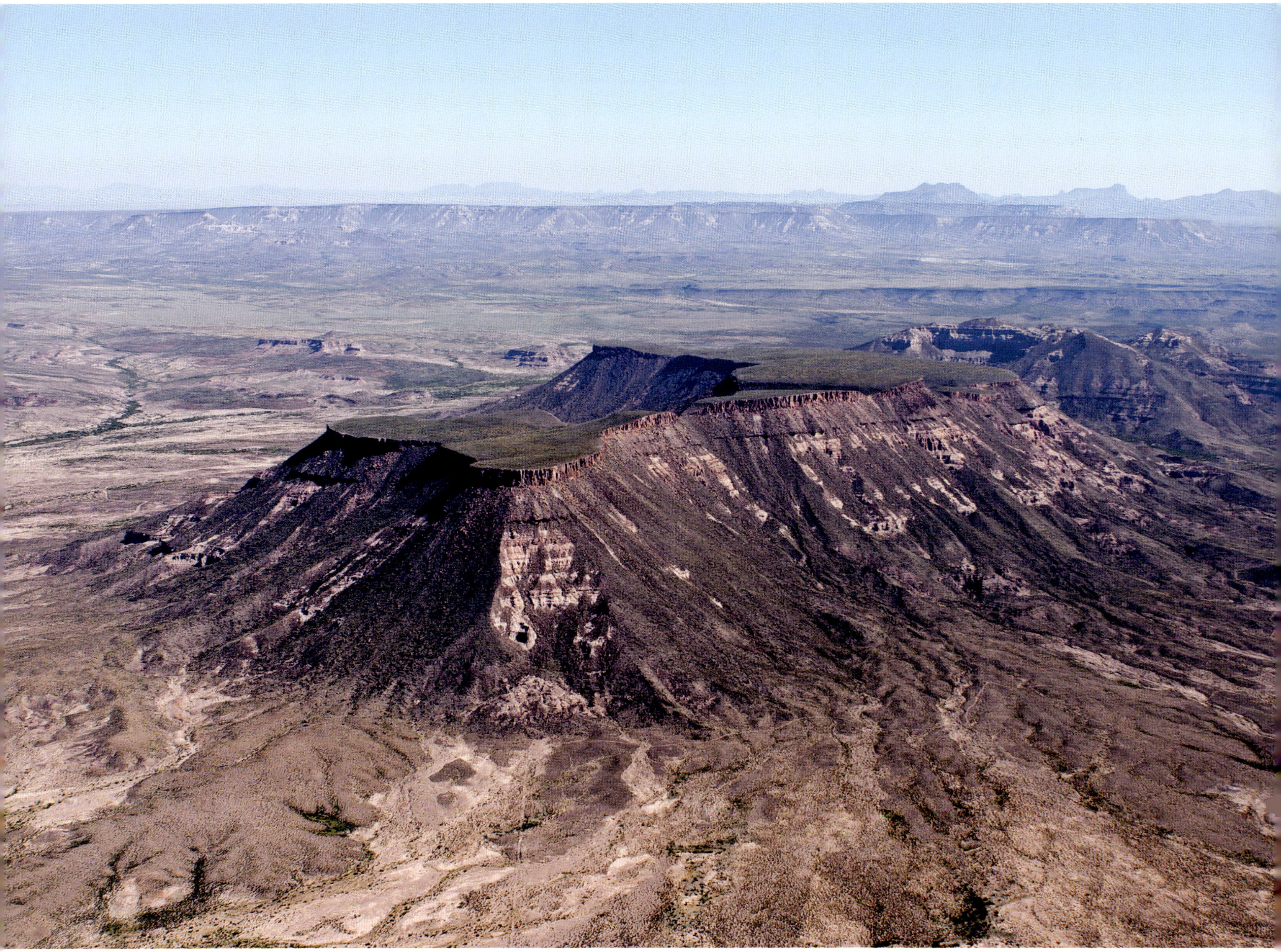

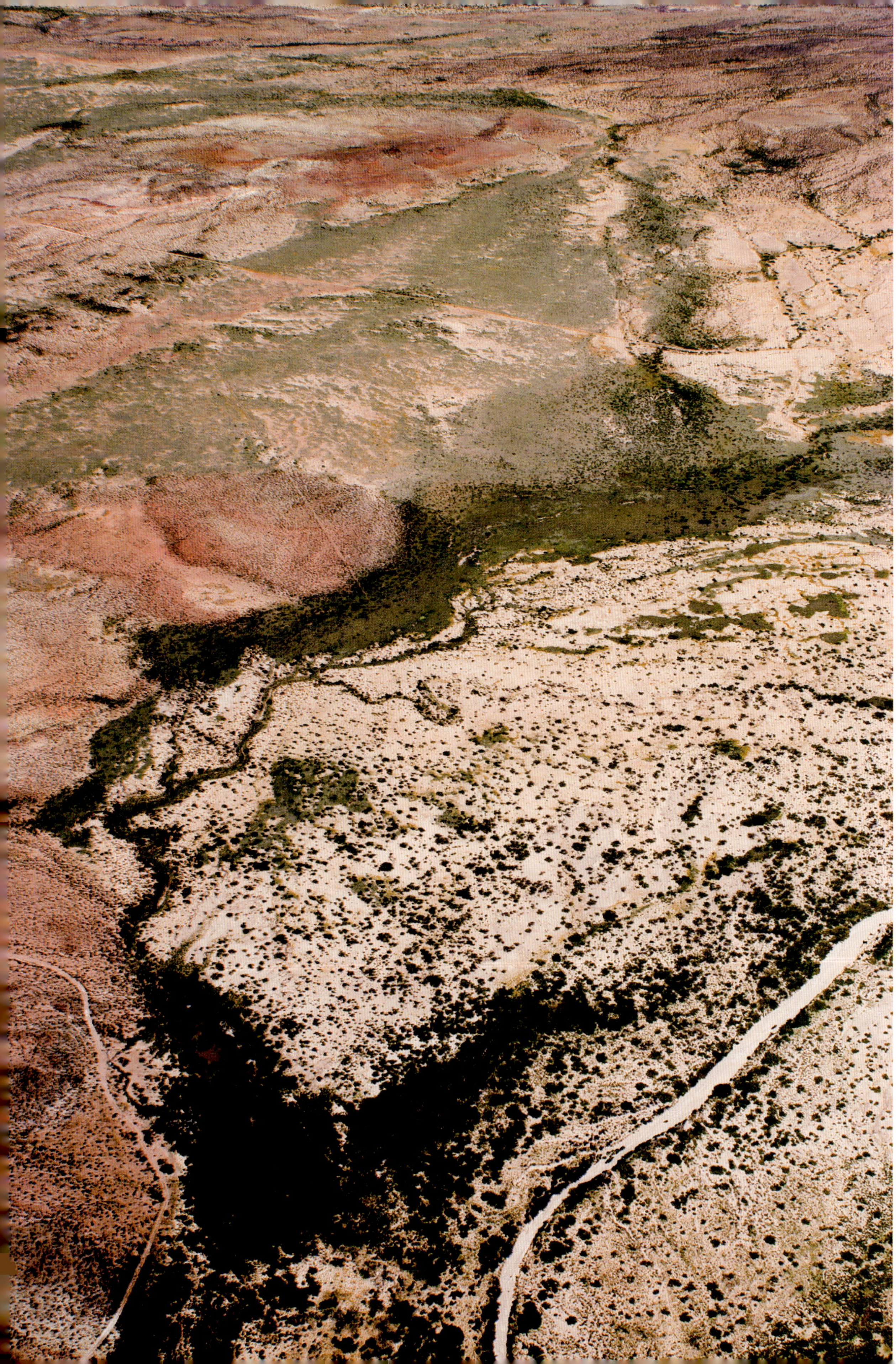

100. Valley Edge Detail

Green Valley

Between Needle Peak and McKinney Mountain, a sash of green hugs the base of the red hills along the edge of Green Valley, creating an impressionistic scene in this austere land. I was charmed by the soft, pastel colors of this scene, especially the earthy reds and the gentle green in the distance.

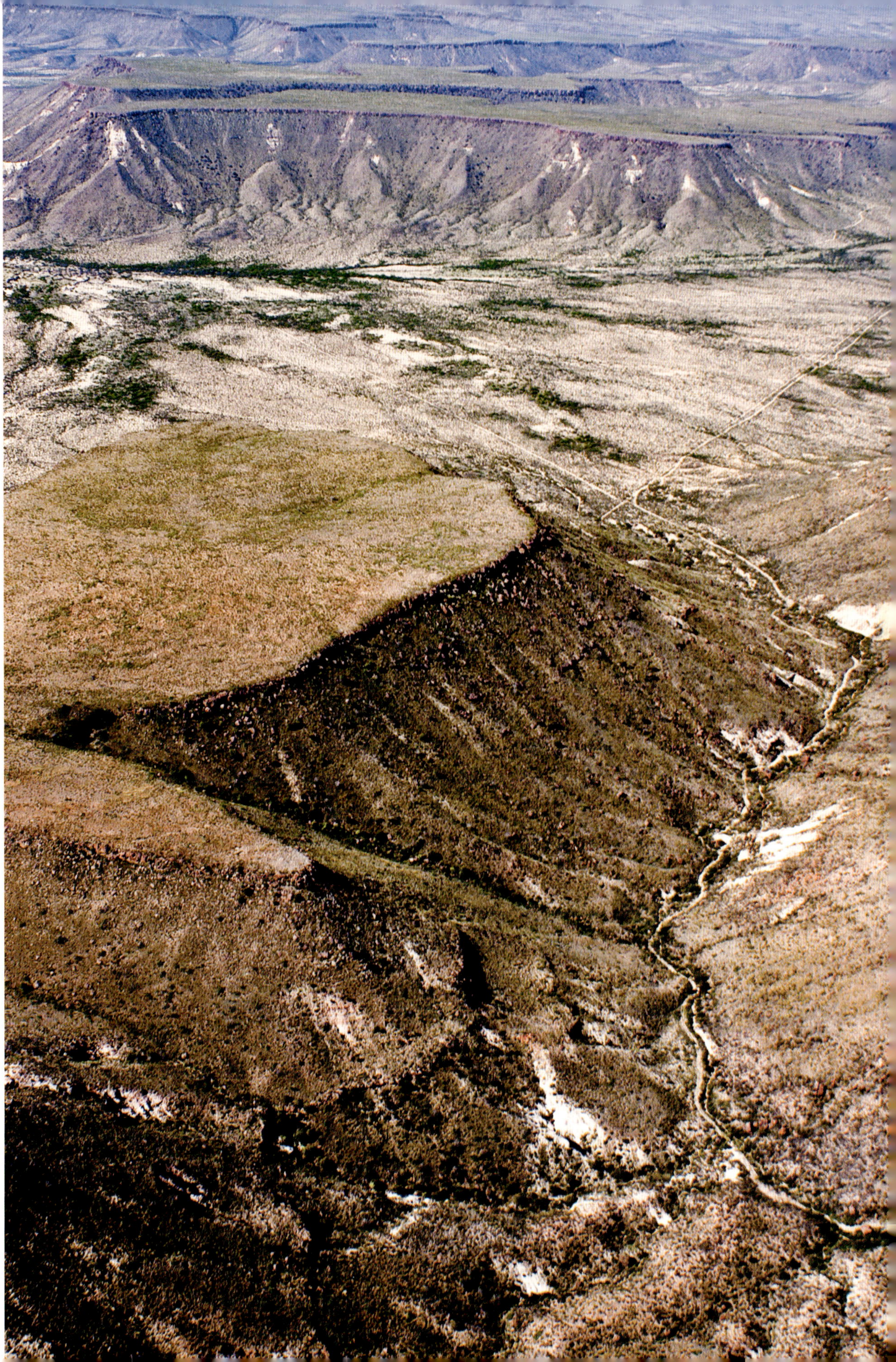

101. Bootlegger Canyon

Twenty-seven miles southwest of Alpine

Located in the Devil's Den topographic map quadrant, Bootlegger Canyon (right) still lives up to its name as a viable path for smugglers coming up from Mexico, then across Green Valley northward to One-O-One Flats. During Prohibition, bootleggers brought illegal Mexican liquor, especially sotol and mescal, across the Rio Grande and north into Texas. In the background (top) is Gemelo Mesa on the north end of Paradise Valley in Presidio County.

X. STEPPING INTO ENCHANTMENT

Crossing the Threshold of the Sierra Vieja to Magical Capote Falls and Candelaria

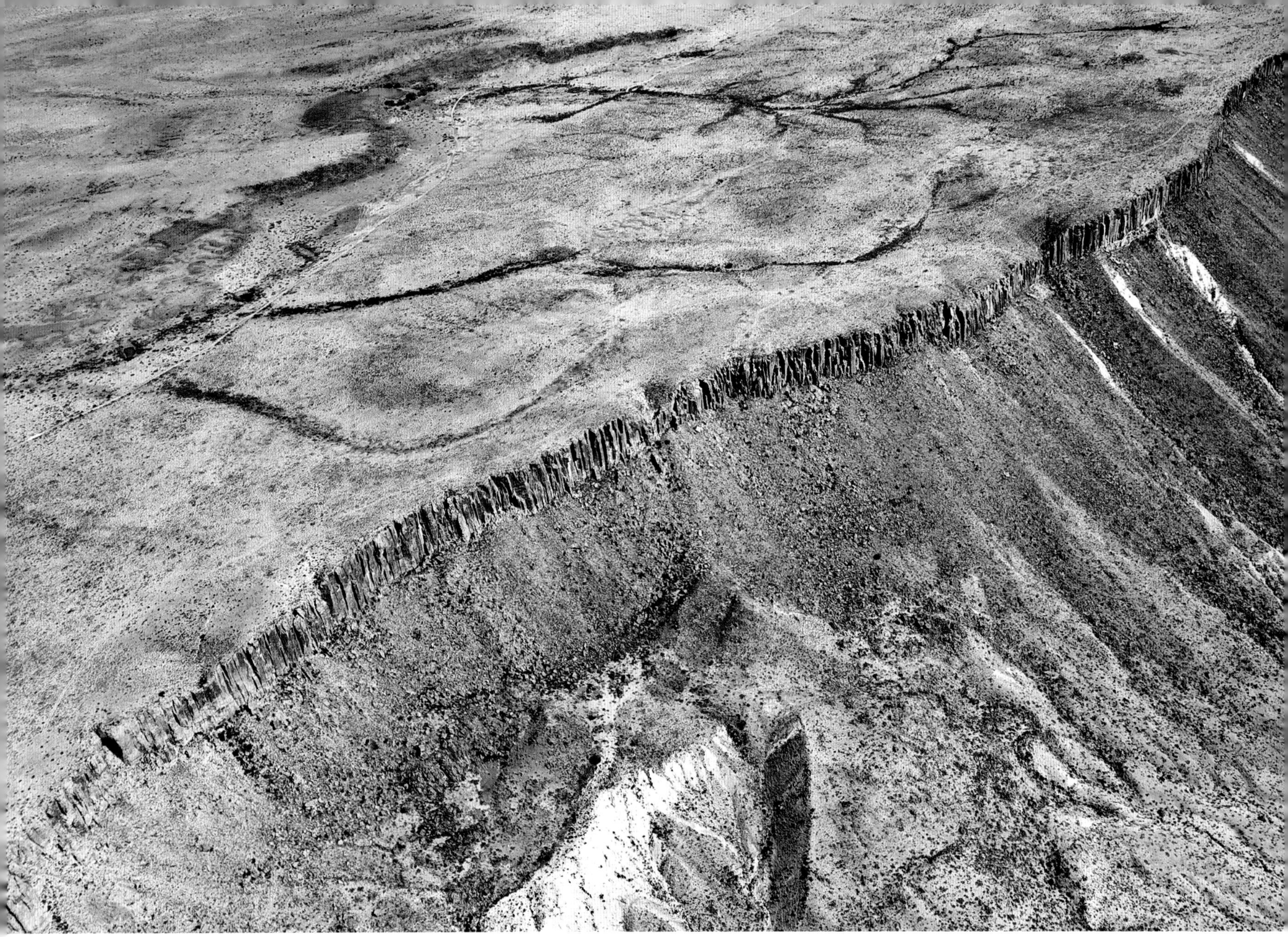

102. Rimrock

Sierra Vieja

The Candelaria rimrock creates an impressive diagonal across this detail of the land west of Capote Peak. The rimrock, known here as the Sierra Vieja, is a formidable natural boundary separating the river valley below from the world to the east. This geologic line of demarcation also divides the grassy range (left) from the arid Chihuahuan Desert (right). The escarpment face consists of tertiary igneous rock. I arranged this photograph to position the cliff as a line extending from corner to opposite corner to emphasize the severe change in topography west of Marfa. I felt I had reached the ragged edge of the flat world.

103. Old Stagecoach Road

Northwest of Capote Falls

Just to the right of center, a white road threads down a break in the escarpment. The Candelaria-Valentine stagecoach used this road on its twice-weekly run. The ability to negotiate the steep drop of the road where it cuts through the Sierra Vieja attests to the skills of the drivers. Legendary Texas lawman O. C. Dowe also used this route to transport wagon loads of building supplies from Marfa when he built his ranch house in the valley below. The layers of this landscape (top to bottom) are the Cienega rim, Cienega Basin, Capote Creek, and the Sierra Vieja rim. Below the Sierra Vieja the land slopes into the Chihuahuan Desert of the Rio Grande valley. This image was one of the first in which color piqued my interest, whereas some others were originally printed in black and white for exhibition.

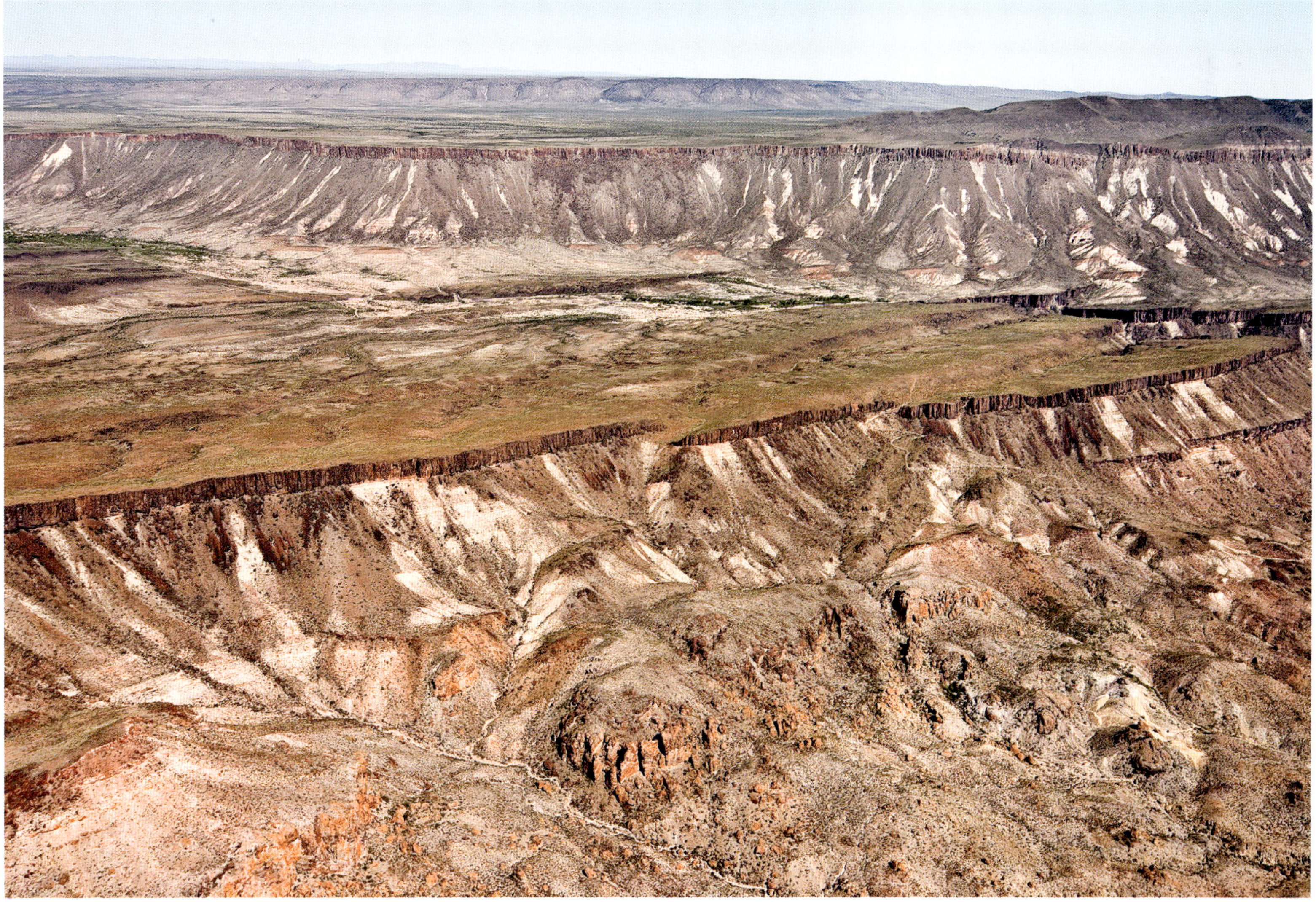

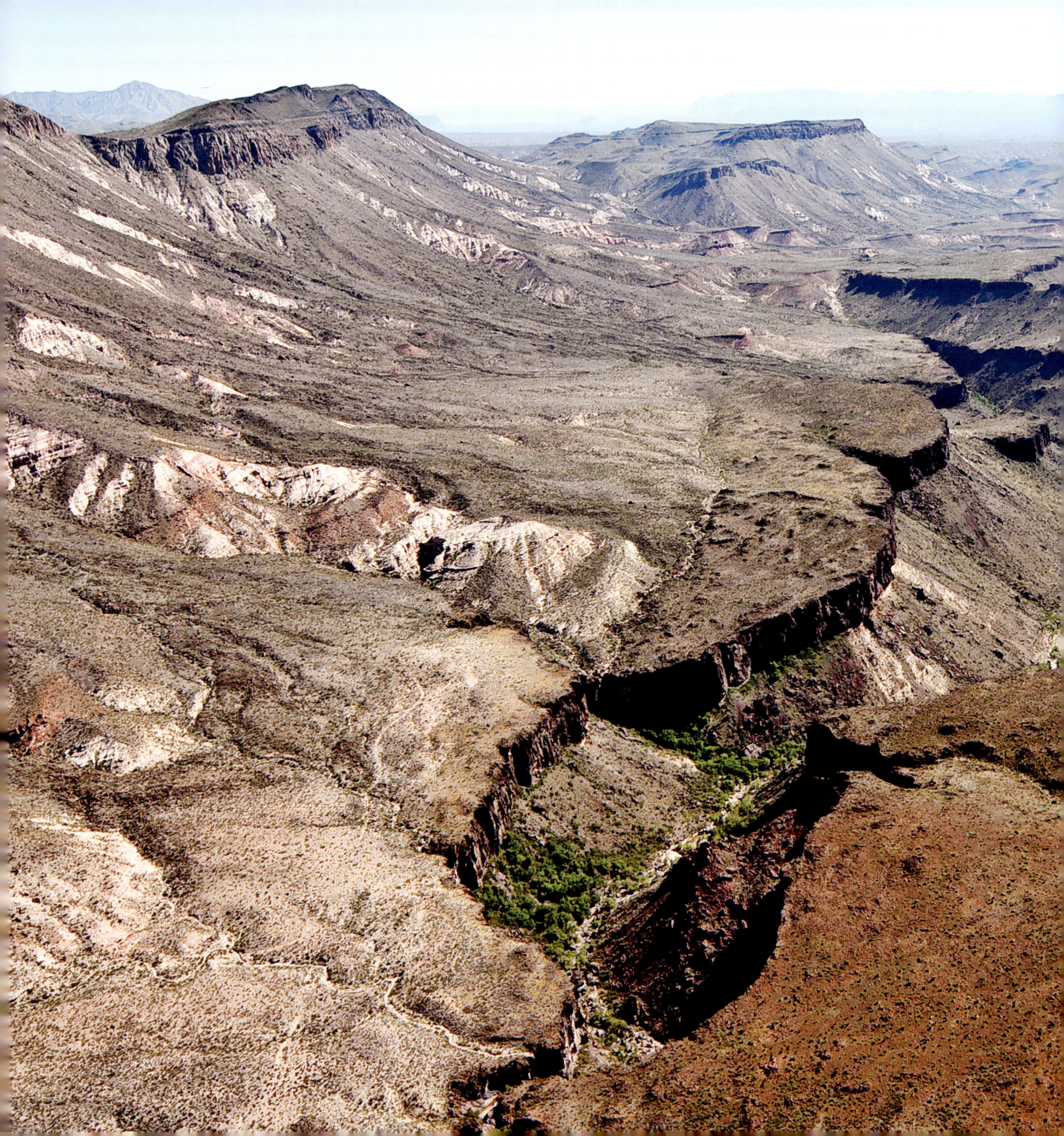

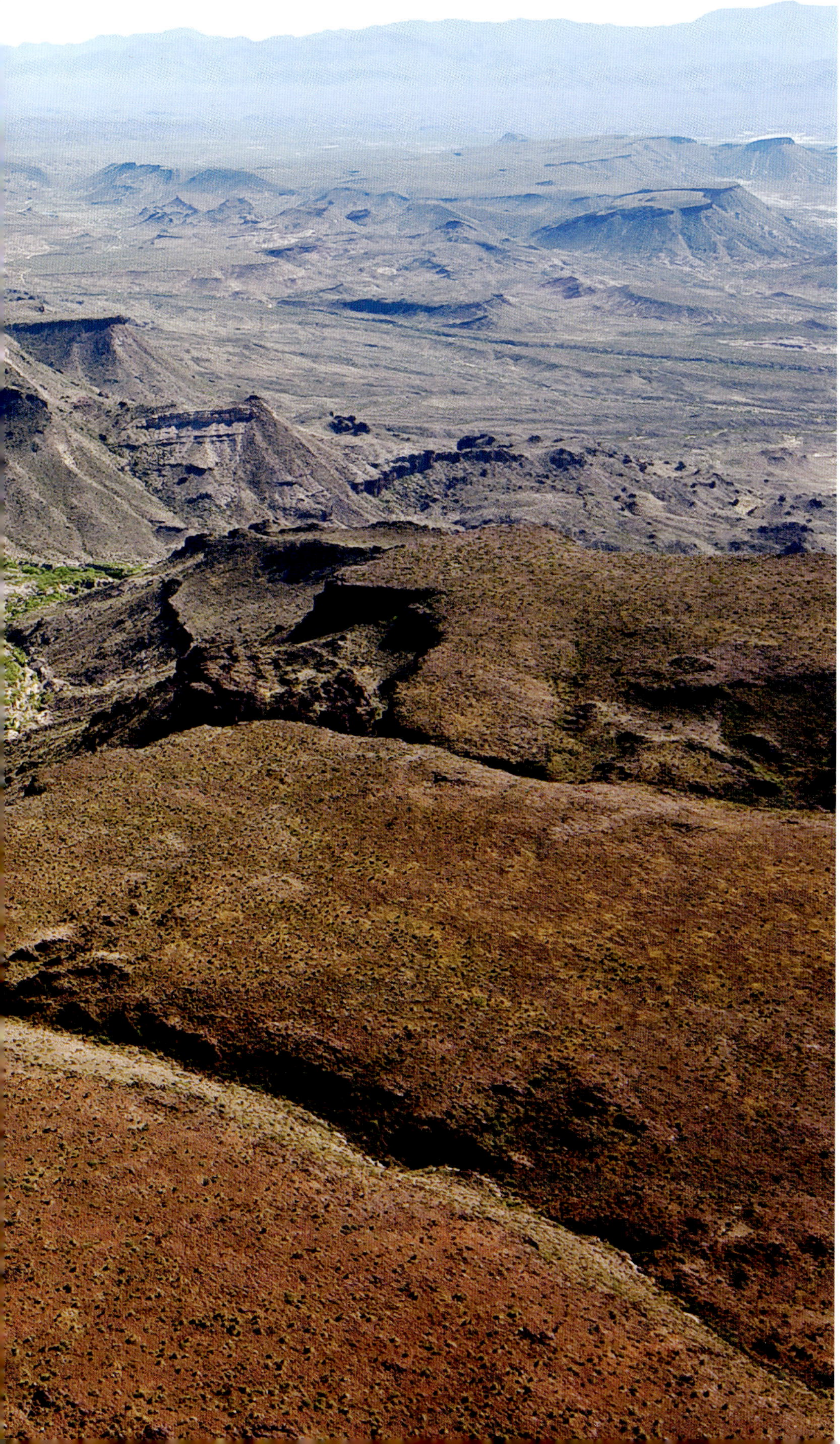

104. Capote Falls

Brite Ranch

Relatively few people have ever seen Capote Falls due to both the forbidding surrounding terrain and location on private land. I was struck by this timeless, almost prehistoric view from the airplane as I looked over the scalloped edges of the eroded hills into the distance. I felt as though a pterodactyl might join us in the flight over the valley.

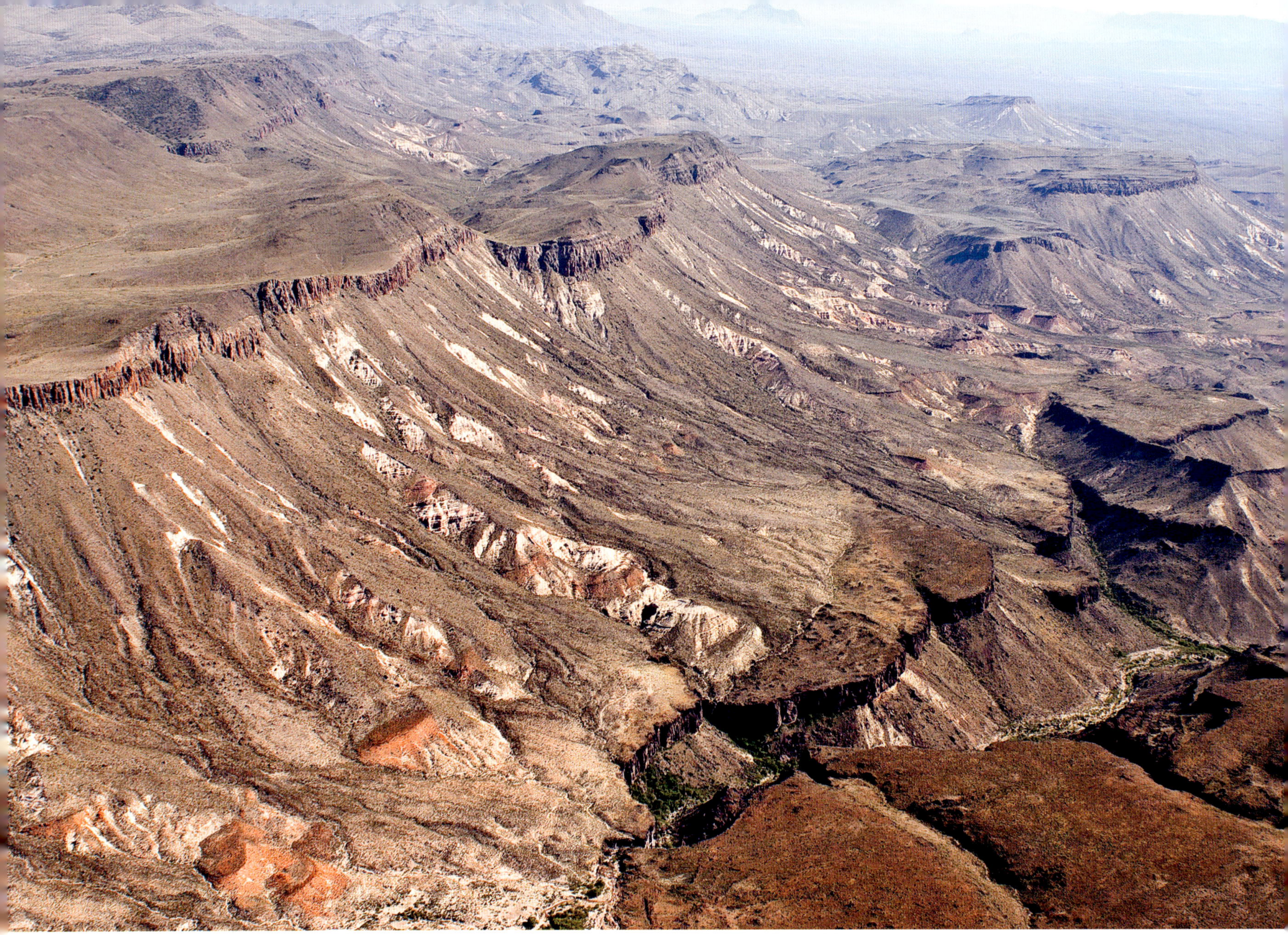

105. El Macho

Thirty-three miles west of Marfa

Looking at these slopes, one can easily understand why the early Spanish explorers never found Capote Falls. Here it seems time has stopped, and before that, it was measured in the ticking of geologic eras. The sculpted hills such as El Macho (just to the left of center) were scooped by erosion and stand looking over the river valley, the thirsty dust of their ashen faces striped only by diminutive Capote Creek, the waters of which evaporate before they ever reach the Rio Grande.

106. The Plunge Pool

Looking upstream

Capote Falls, 175 feet high (top, center), is the highest continuously flowing waterfall in Texas. The water source begins in a marshy spring, or *cienega* in Spanish, at the base of Capote Peak, then flows three miles to the falls, where it plummets into the plunge pool. The box canyon below cuts through the Sierra Vieja to a depth of about two hundred feet. The downstream mouth of Capote Canyon has a failed nitrate mine, dating from about 1918, in the lower wall and rusting remnants in the bottoms. At the left, the geology of the area is revealed: the hard upper crust is rhyolite, with the softer slopes being volcanic tuff. The riparian zone created by Capote Falls is home to numerous birds and plant species, some rare.

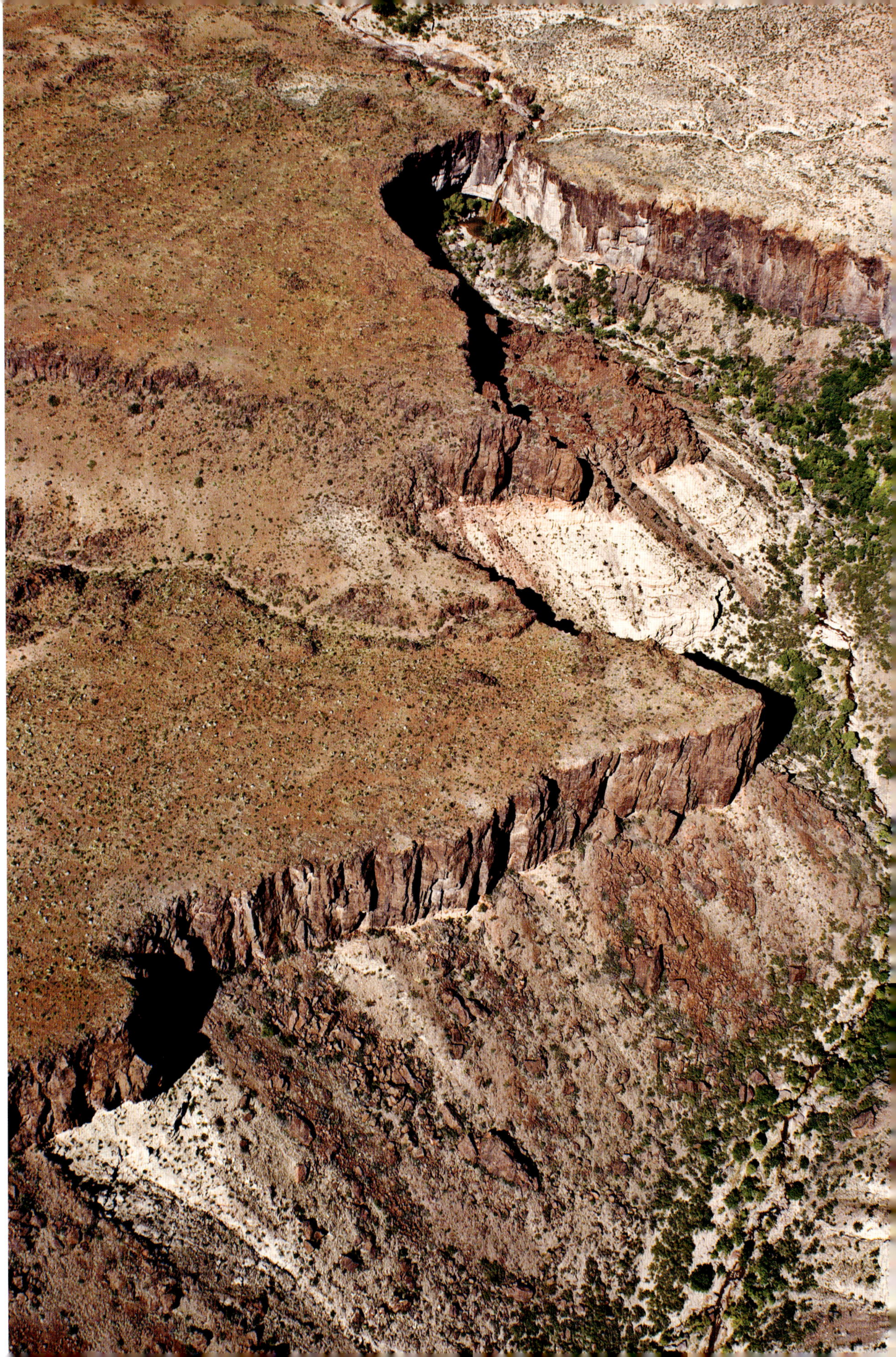

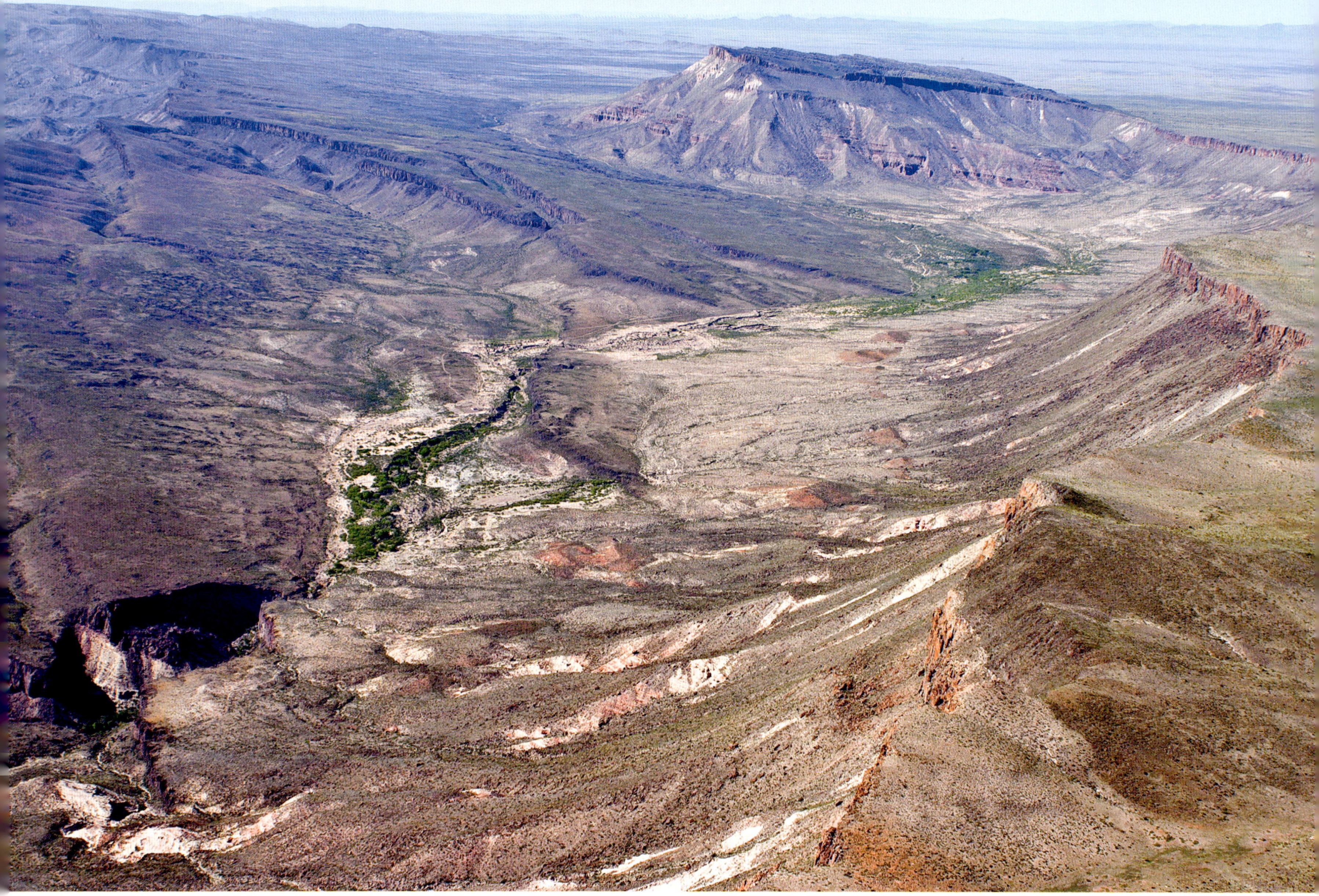

107. The End of the Trail

Capote Peak

Formidable Capote Peak (center) rises 6,212 feet above this lonesome land that exists in its near-pristine state. In the shadow of the peak is the headquarters of working Brite Ranch, founded by L. C. "Luke" Brite, who registered the Bar-Cross brand in 1904. For fourteen years Brite sold a thousand bulls per year. The ranch's line-bred Hereford cattle are considered superior and well suited for the desert Southwest. Today the ranch is run by one of L. C. Brite's fourth-generation grandsons, Jim White III, who continues the family's excellent range management and business practices. The ranch still has traditional roundups and brandings—although the chuck wagon has been replaced by a vintage "chuck truck." I was awestruck by the painterly feel of this vista and instinctively composed the image to have the creek and ridge lead the eye of the viewer into the distance toward the mountain and then beyond to help create an experience of vast space.

108. Painted Chihuahuan Desert Canyon

Twelve miles east-northeast of Ruidosa

Just below the Cleveland Flat (top) and headquarters of the Petan Ranch (upper center), Hot Springs Creek gouges a parfait-colored canyon as it wanders west toward the Rio Grande. About three miles to the west is Chinati Hot Springs, which offers a relaxing geothermal hot spring soaking experience with solitude and air-conditioned cabins.

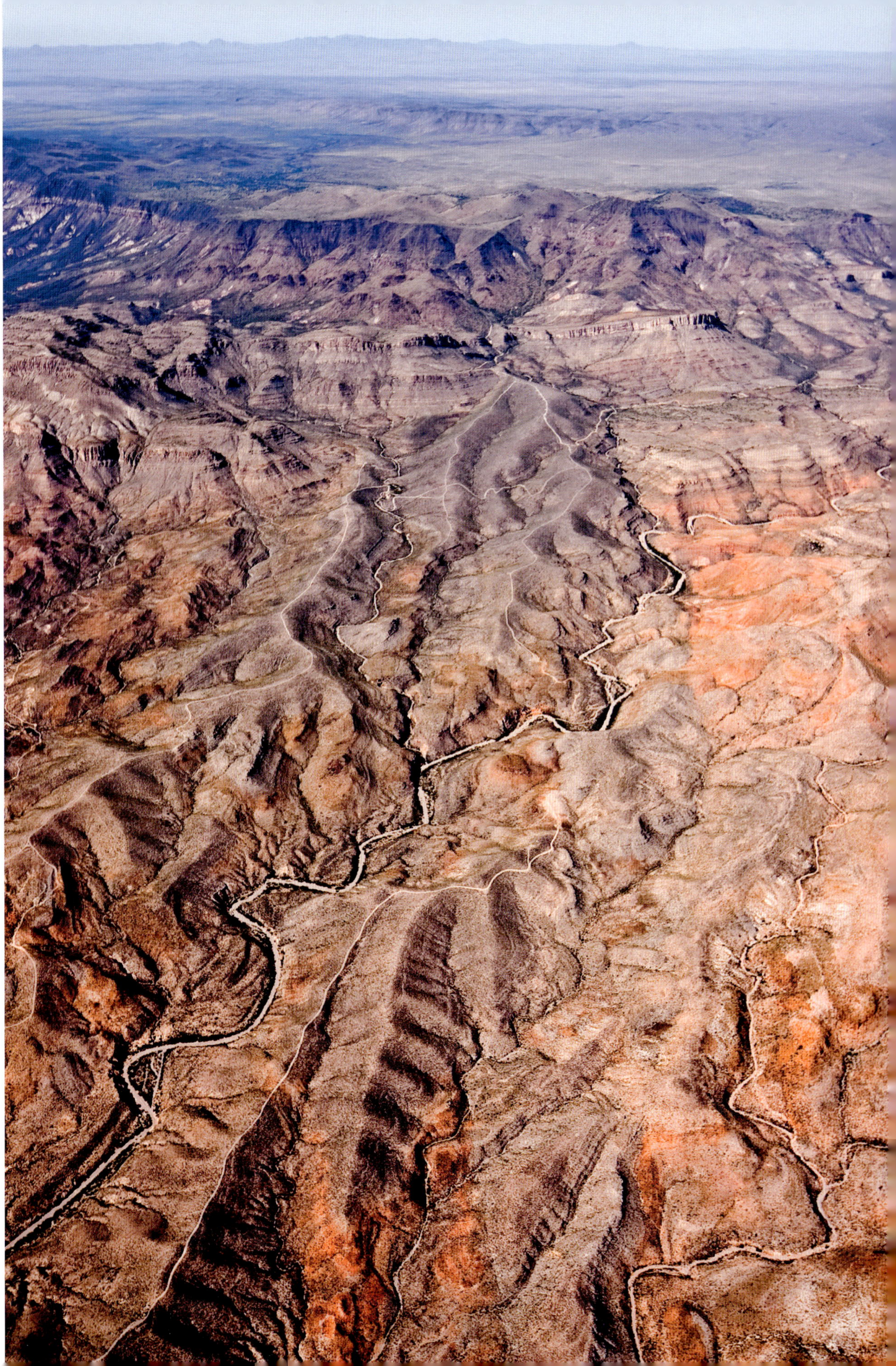

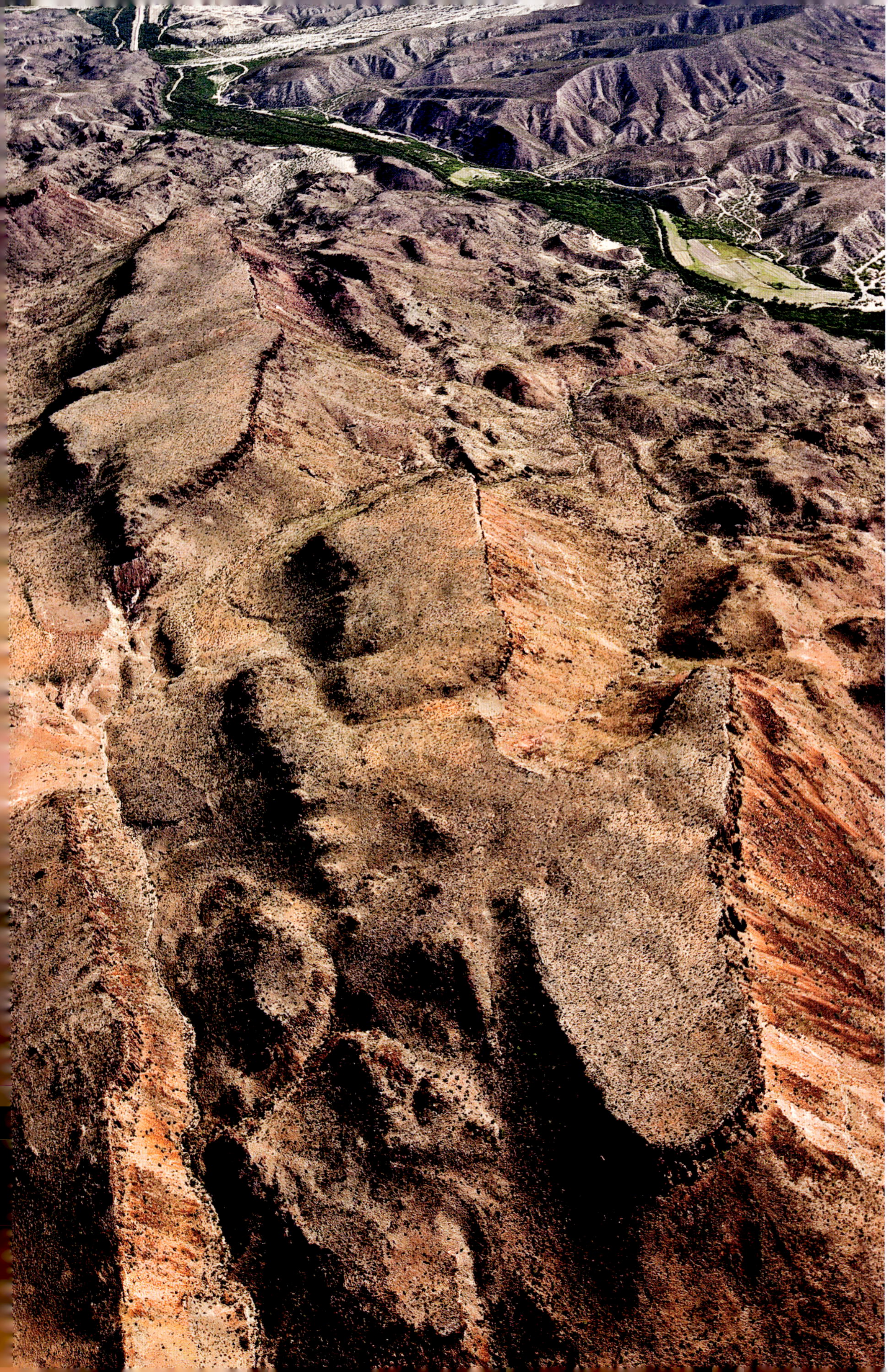

109. Looking toward Cerros Colorados, Mexico

Upstream from Candelaria

Lozenge-shaped mesas up to 3,932 feet in elevation line the US side of the Rio Grande. Along the Mexican side of the river (left to right) are Baeza Crossing, the farmlands of tiny Cerros Colorados village, Arroyo El Alambre, and San Antonio Crossing. This area has numerous unofficial international river crossings that people on both sides freely traversed. No doubt there is smuggling; yet people also visited friends and family, worked, bartered, bought supplies, and sometimes sought emergency medical care across the river. The US government closed the crossings, citing concerns including terrorism. Vigilant locals disagree, saying that they have never seen a terrorist, and closings have hurt the local economy and caused personal hardships. The phantasmic shapes of the mesas in the foreground immediately inspired me to take this frame. To give a reference for scale, I included the Rio Grande in the background.

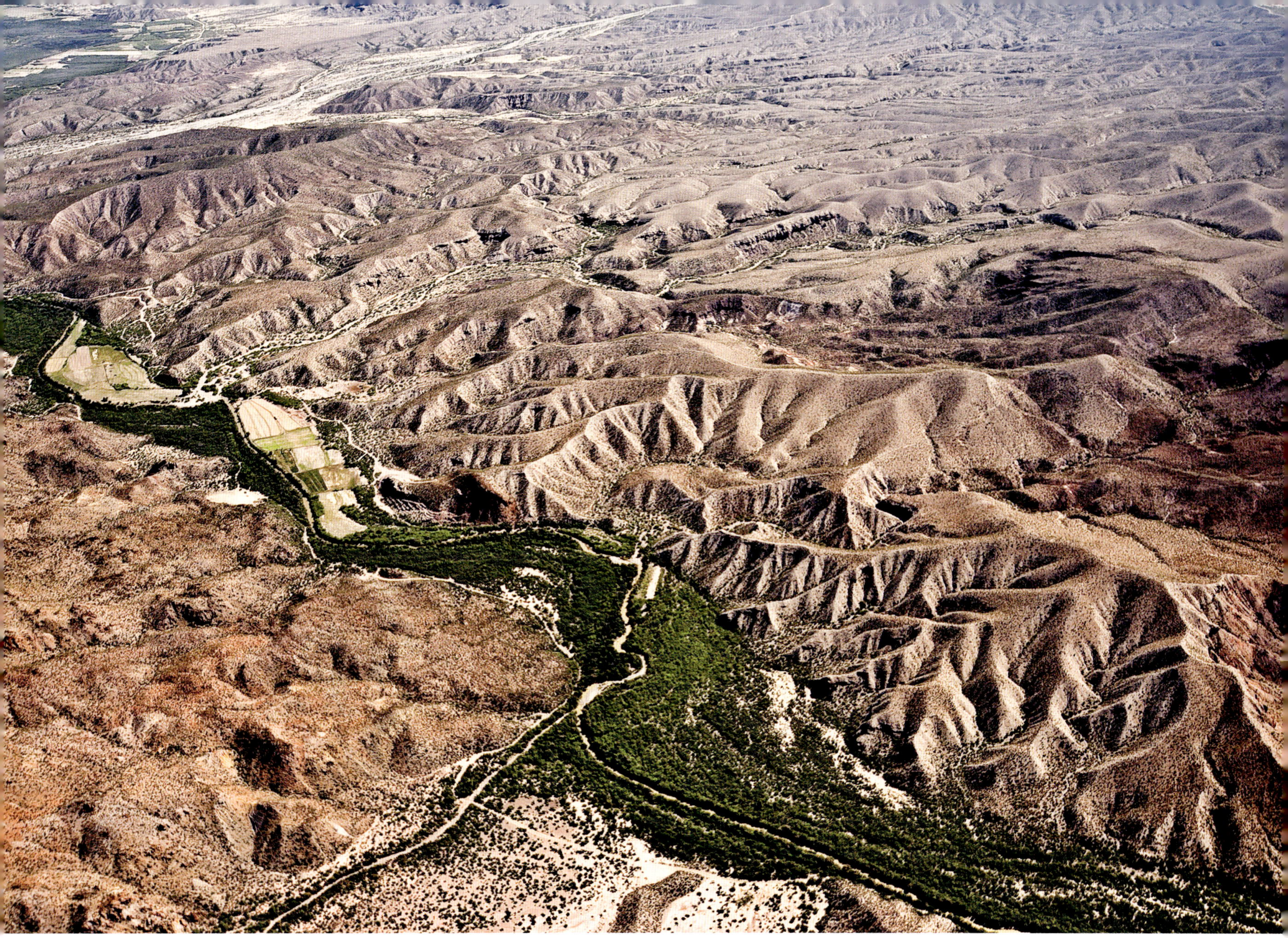

110. Silver Mountains

Rancho Baeza

Looking downstream toward Las Palomas, Mexico. The Cerros Colorados mountains shimmer like brushed silver in the Chihuahuan Desert sun. On the Mexican side, Arroyo La Jarrilla cuts between the two green patches (right to left) of Rancho Baeza and Cerros Colorados village. On the US side, Chispa Road, more of a jeep trail, hugs the base of the hills and is joined by an unnamed ranch road (lower left) that turns inland.

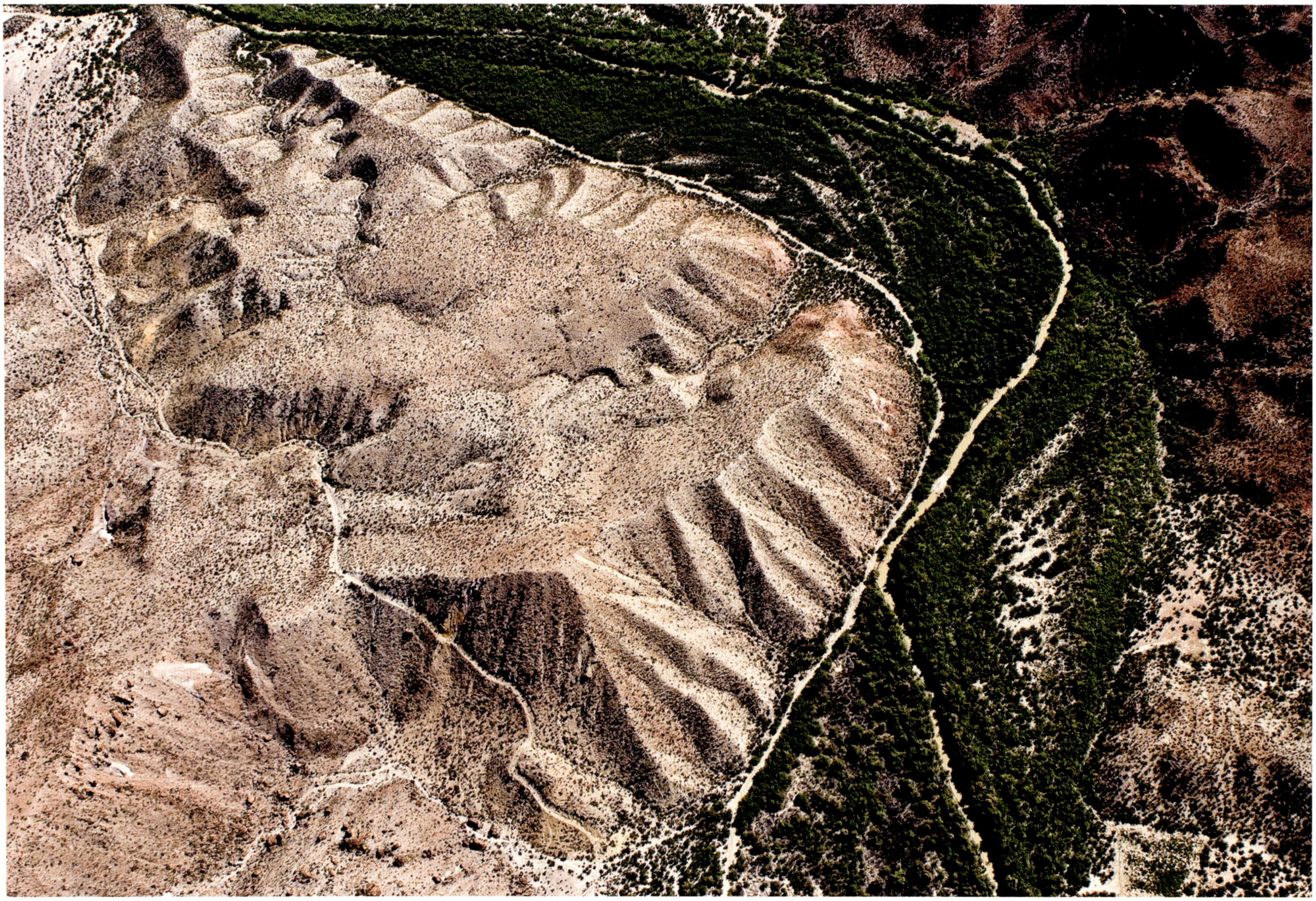

111. The Road to Chispa

Along the Rio Grande

The wetlands along the Rio Grande are home to wildlife including deer, javelina, ducks, geese, and heron. Bird watchers with four-wheel-drive vehicles can find many species in this less visited part of the Rio Grande riparian zone that is a preferred route for spring and fall migrations. Fall is the most exciting season for birding. The floodplain is infested with salt cedar, a nonnative plant that has proliferated to the point that it is impacting the river ecosystem. A great concern is that the plant secretes salt, which kills native plants and shrubs. Experts have described the invasive water-hungry salt cedar as one of the greatest ecological disasters in the history of the Rio Grande.

112. Candelaria

Rio Grande looking south

In the distance is the Texas village of Candelaria (left). It takes its name from candelilla, the desert plant that was used to make wax, most notably for waterproofing tents during World War I. The plants were harvested from the wild and then boiled to release the wax. During the Mexican Revolution, Candelaria was the location of a US Army camp and a crossing point to pursue raiders into Mexico. It is also the spot where Capt. Leonard F. Matlack crossed the Rio Grande in 1919 to ransom two Army Air Service "river pilots" who crash-landed their biplane in Mexico and were taken hostage by Jesus Renteria's gang. Today Candelaria sits near the ruins of a Rio Grande footbridge that was destroyed by the US Border Patrol in 2008. On the Mexican side, the villages (bottom to top) are Cerros Colorados, Las Palomas, Las Cienegas, San Antonio el Bravo, and La Oficina. Looking down the Rio Grande from the heights shown in this photographic vista still gives me butterflies.

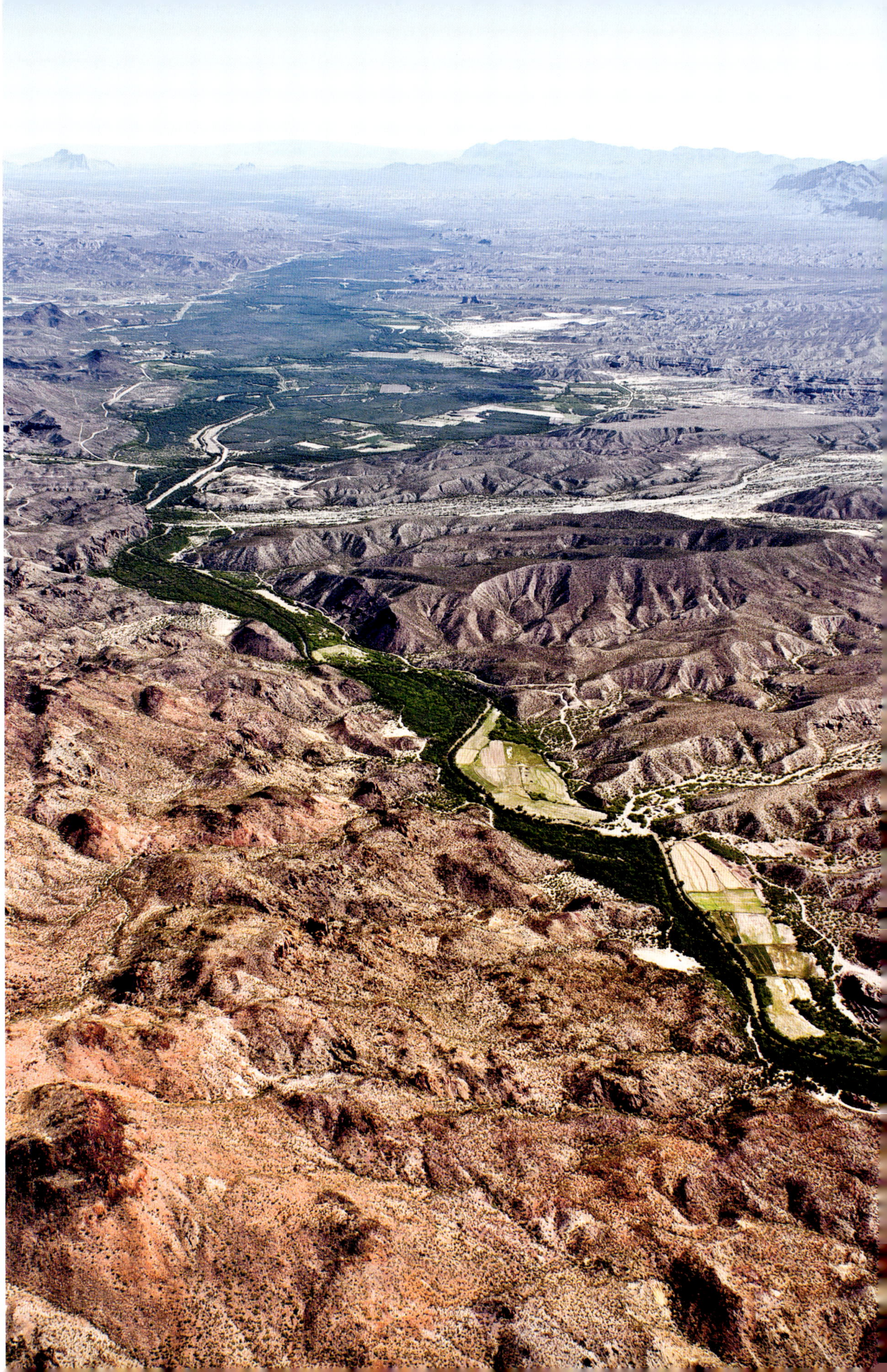

XI. UP THE STARRY STAIRWAY

Exploring the Pristine Davis Mountains and McDonald Observatory

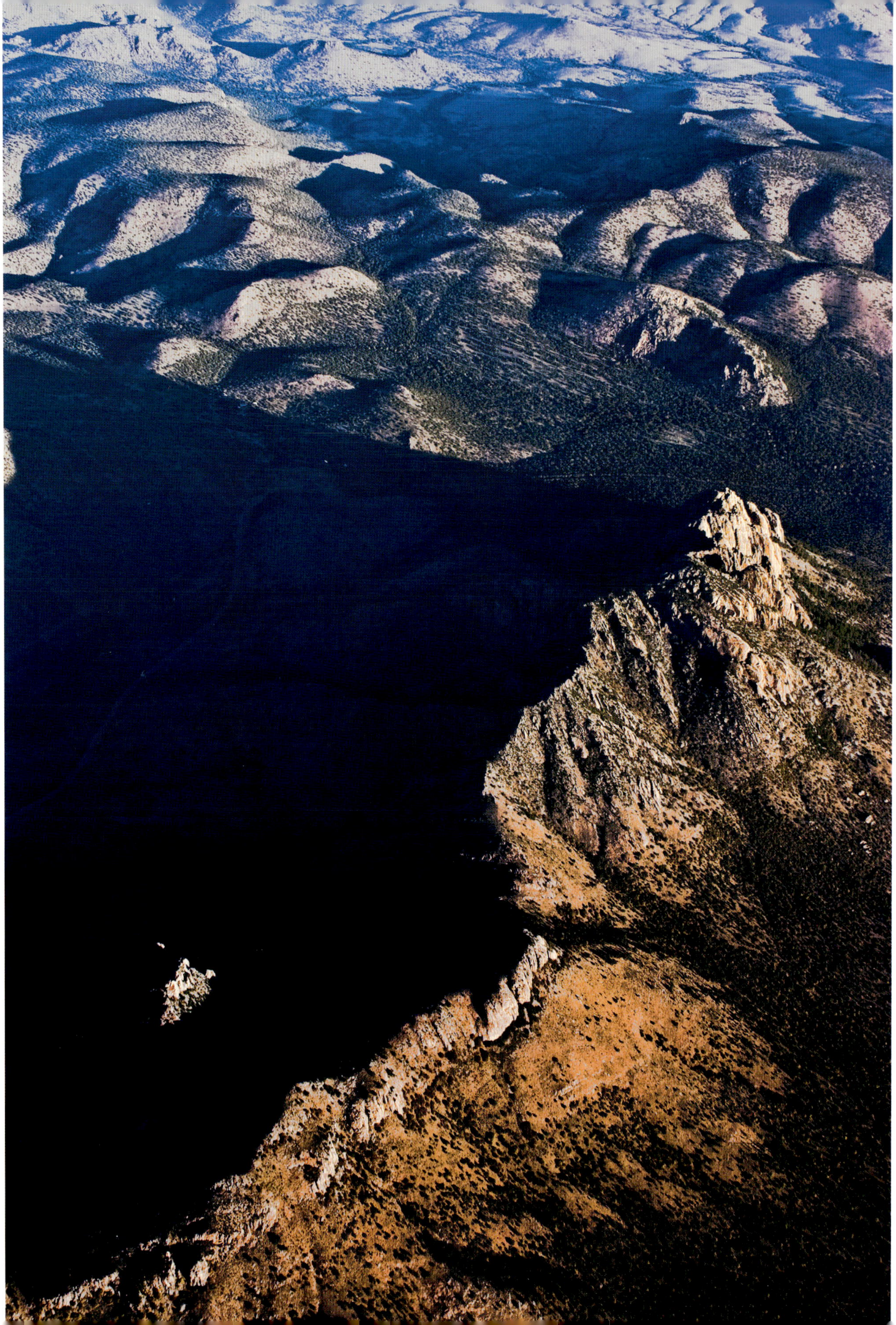

113. Sawtooth Mountain, Palisade Detail
Twenty-eight miles north-northwest of Marfa
The sheer face of the Sawtooth Mountain palisade soars over Broke Tank Draw (upper center). Looking down the spine of this ridge, one can see how ancient magma pushed up into a giant crack in the earth that served as a mold for the mountain. Later erosion washed the more vulnerable material away, and dramatic hard rock of the sill remained. The peak is 7,686 feet high.

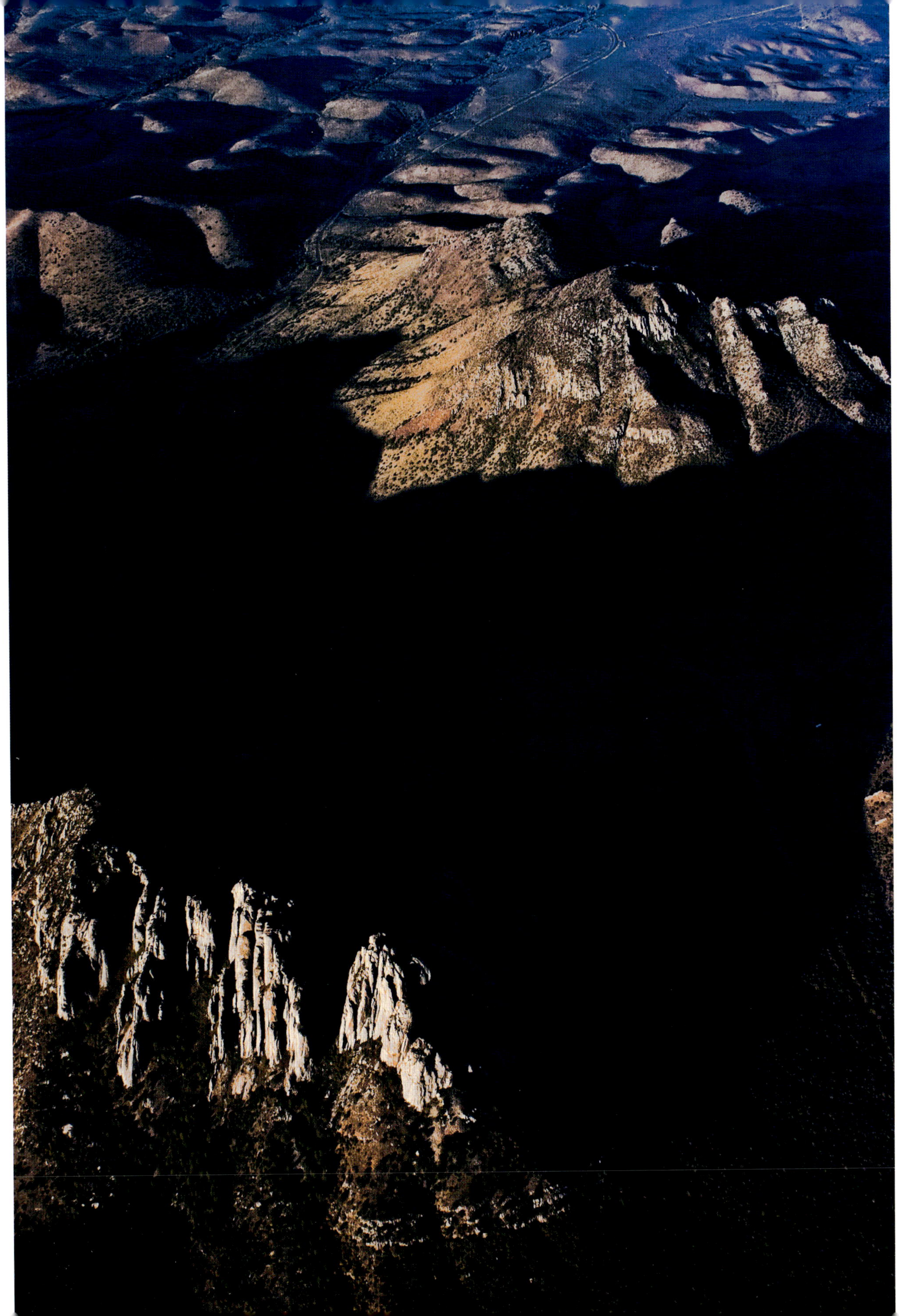

114. Sawtooth Mountain, Columns Detail

Southeast of Texas Highway 116

Projecting out of the shadows, Sawtooth Mountain shows its namesake rocky fangs. The columns are made of igneous syenite that is hard as granite. Texas Highway 116 passes through the valley in shadow above the spires. Bear Mountain (upper right), elevation 7,256 feet, seems tame by comparison. My motivation to pair these two images was the riveting negative space between the mountains and how those shadow areas interact when the two images are placed side by side as a diptych.

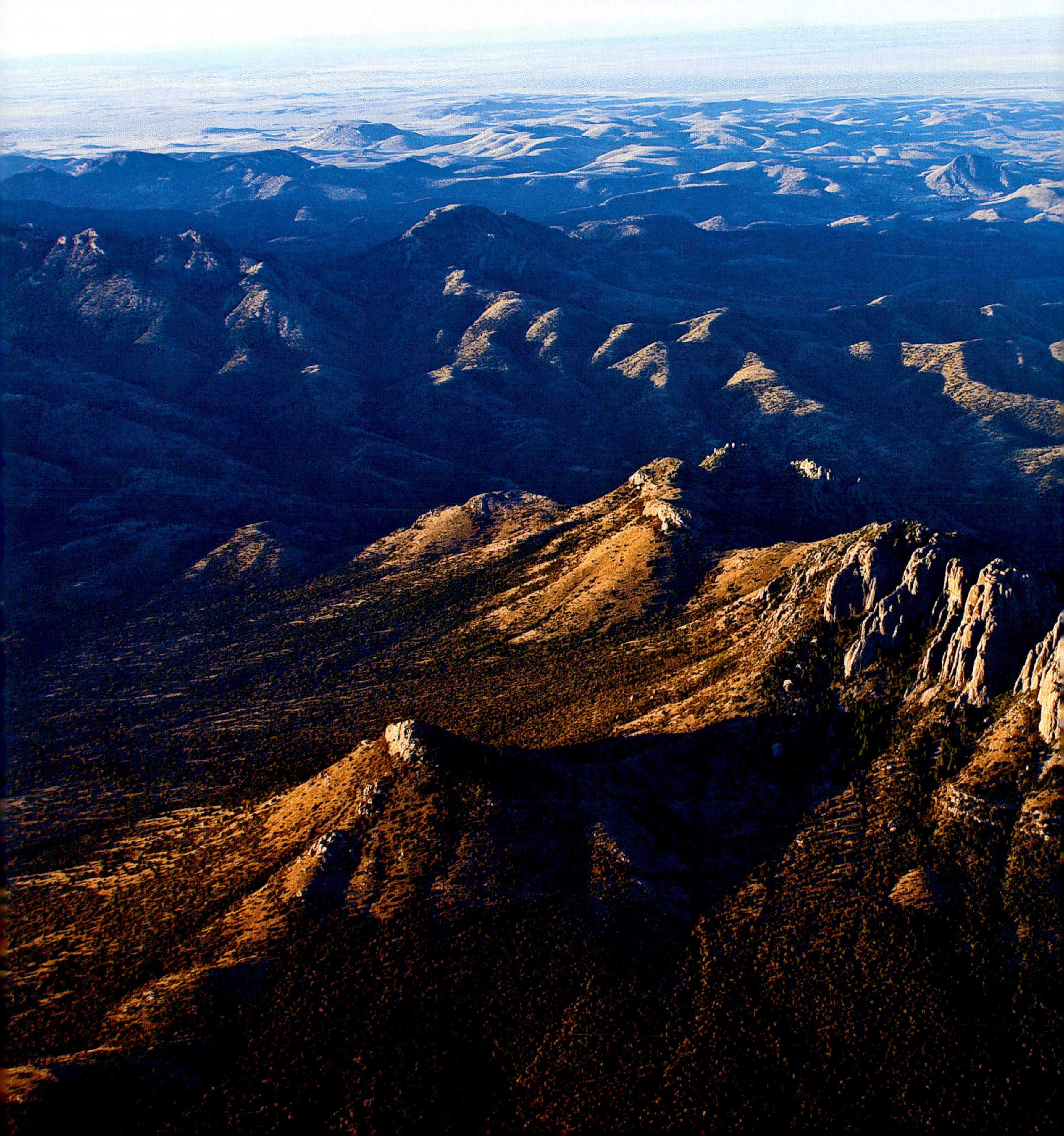

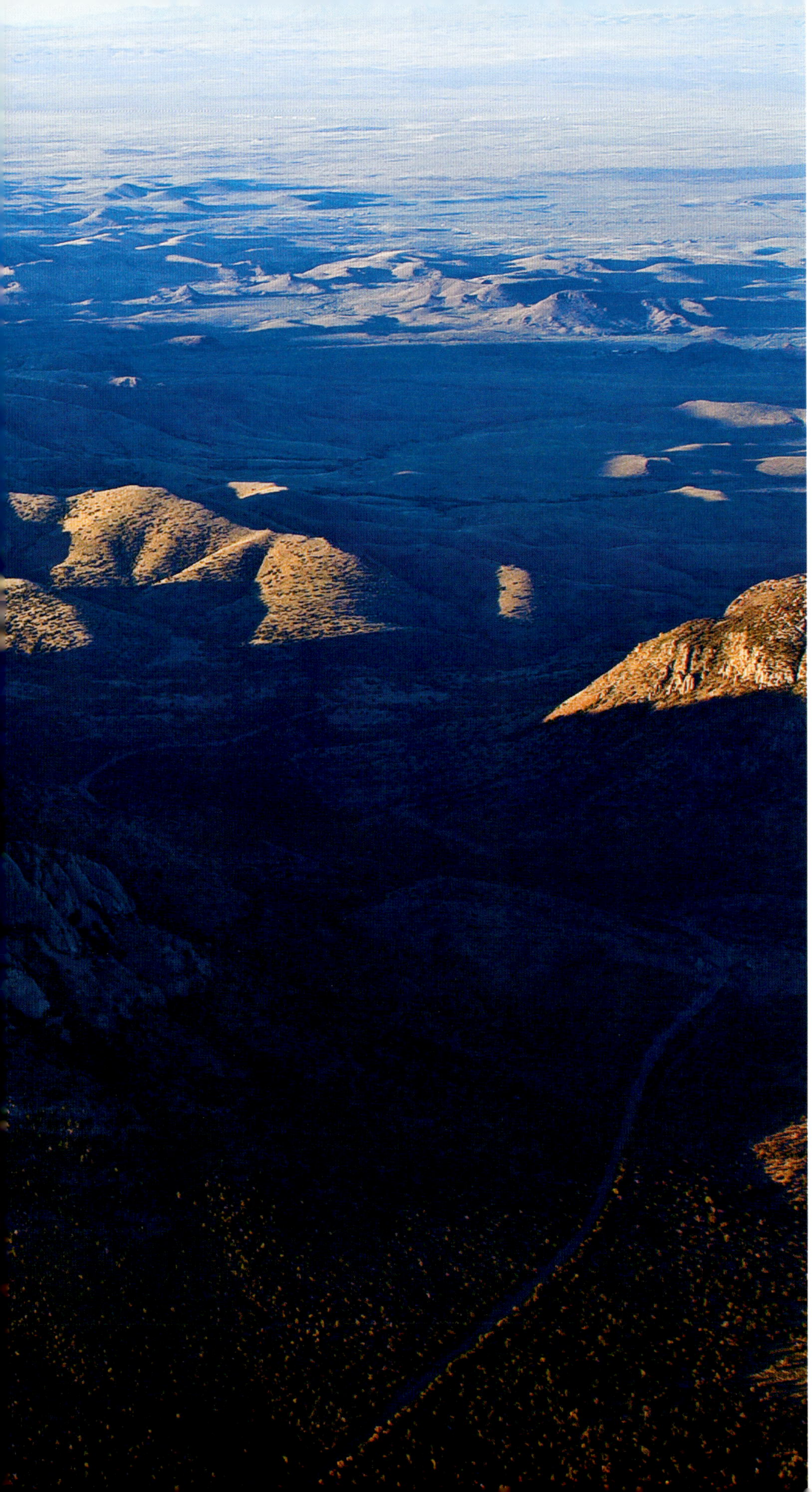

115. Davis Mountains Vista

Twenty-one miles west-northwest of Fort Davis

The Davis Mountains, covering more that 600 square miles, make up the largest mountain range that lies wholly within the state of Texas. The mountains are named for Jefferson Davis, the US secretary of war under President Franklin Pierce and later the president of the Confederate States of America during the Civil War. Jeff Davis County, which covers this area, and Fort Davis, which is the county seat, are also his namesakes.

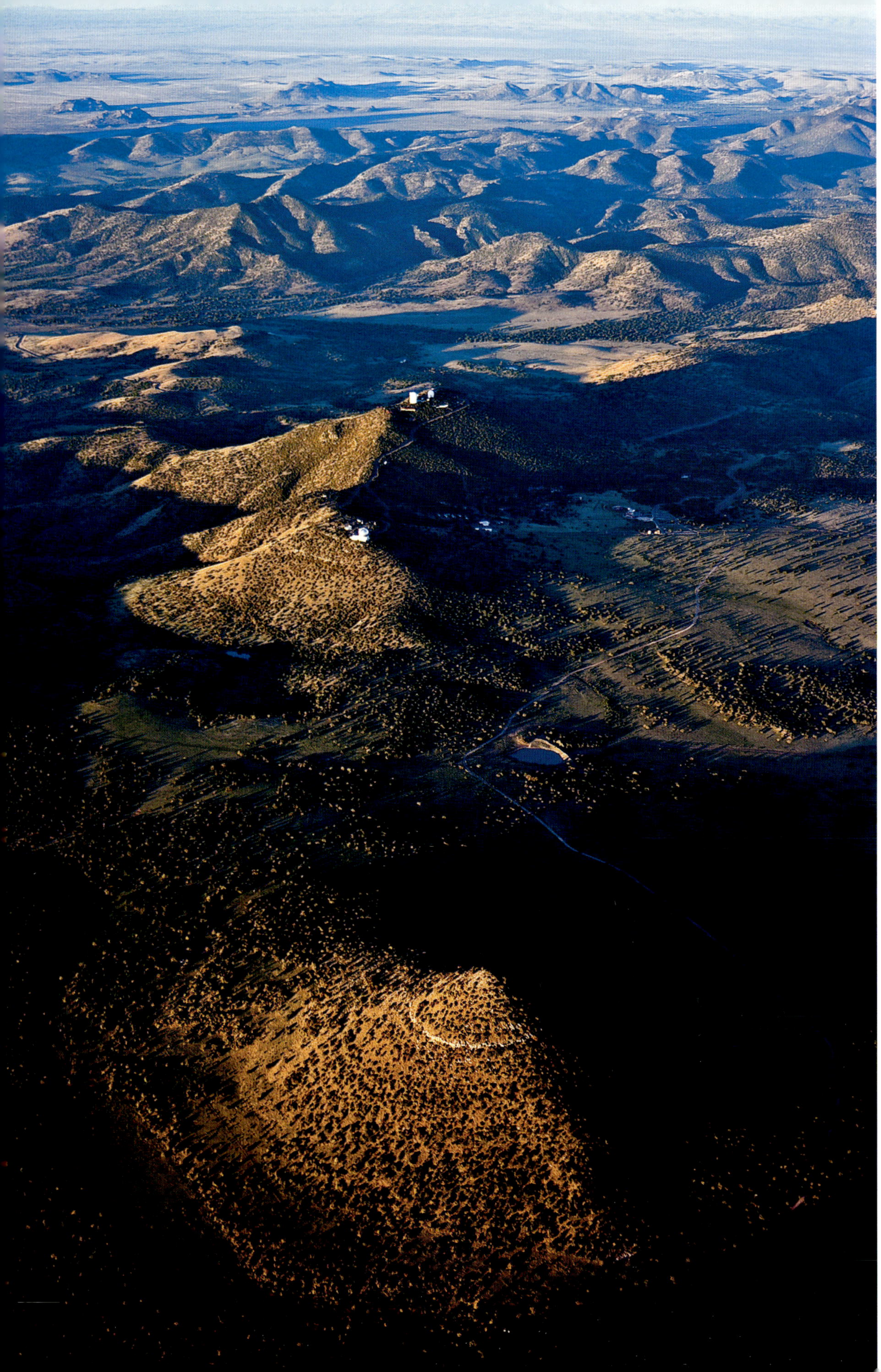

116. Telescope Domes

McDonald Observatory

The large telescopes of McDonald Observatory are the Hobby-Eberly Telescope on Mount Fowlkes (lower center) and the Otto Struve Telescope and the Harlan J. Smith Telescope on Mount Locke (upper center). As a research unit of the University of Texas at Austin, the observatory uses an array of telescopes and instruments to do research on planets, stars, galaxies, and theoretical astronomy, including dark energy. McDonald Observatory is open to the public with a wonderful public outreach program including educational presentations and stargazing parties.

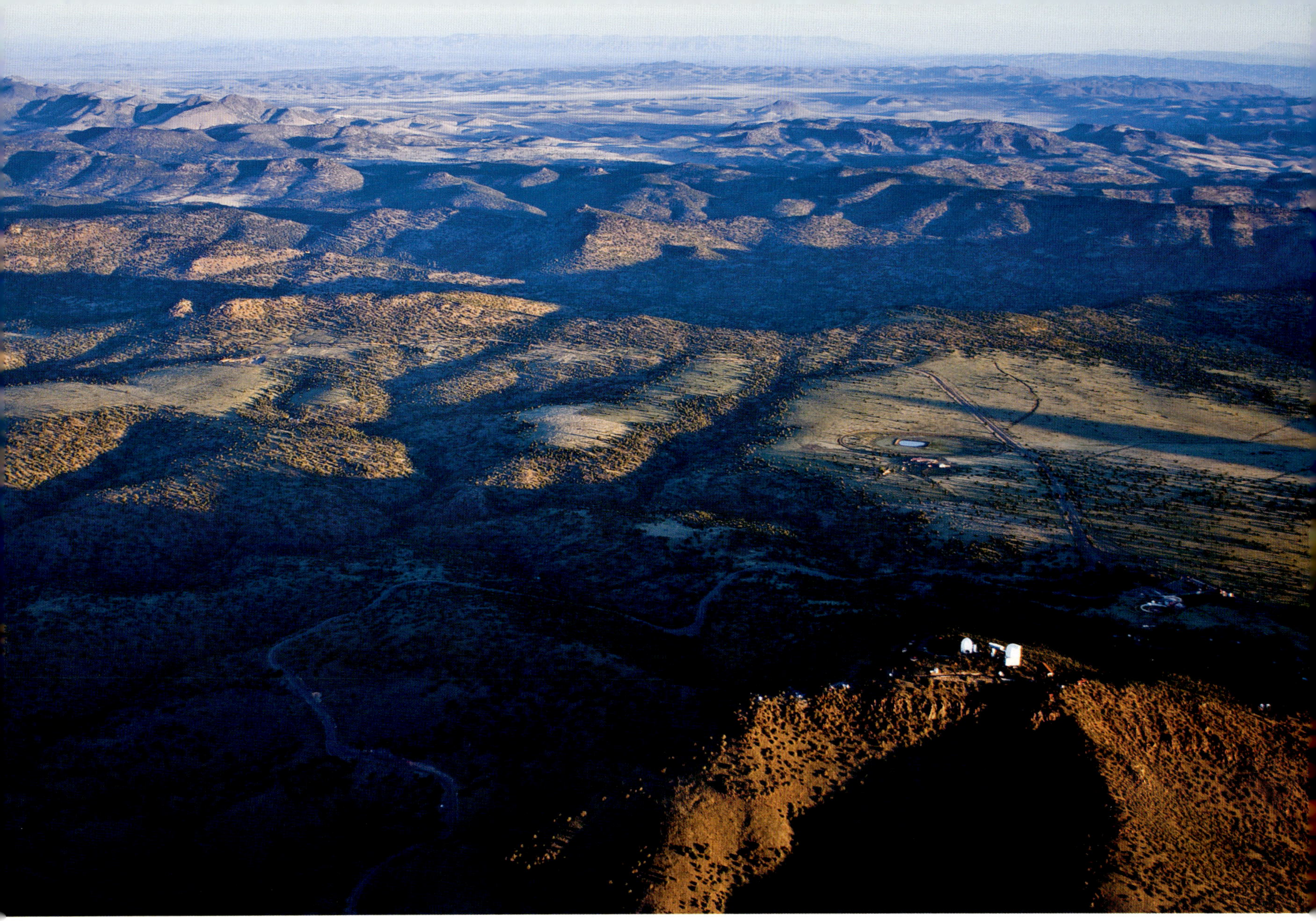

117. The Highest Paved Road in Texas

Ten miles northwest of Fort Davis

Set in the largely undeveloped Davis Mountains, McDonald Observatory benefits from the Dark Skies initiative to reduce nighttime light pollution that could interfere with the observatory's work. Texas Spur 78 climbs nearly to the top of Mount Locke, making it the highest paved road in Texas at 6,791 feet elevation. Recently the observatory's Frank N. Bash Visitors Center dedicated the Wren-Marcario wheelchair-accessible telescope, which is the first of its kind. Just beyond the observatory, back on earth, is the Eppenauer Ranch with its oval horse-training track. The Nature Conservancy has been acquiring large amounts of property for conservation in the Davis Mountains.

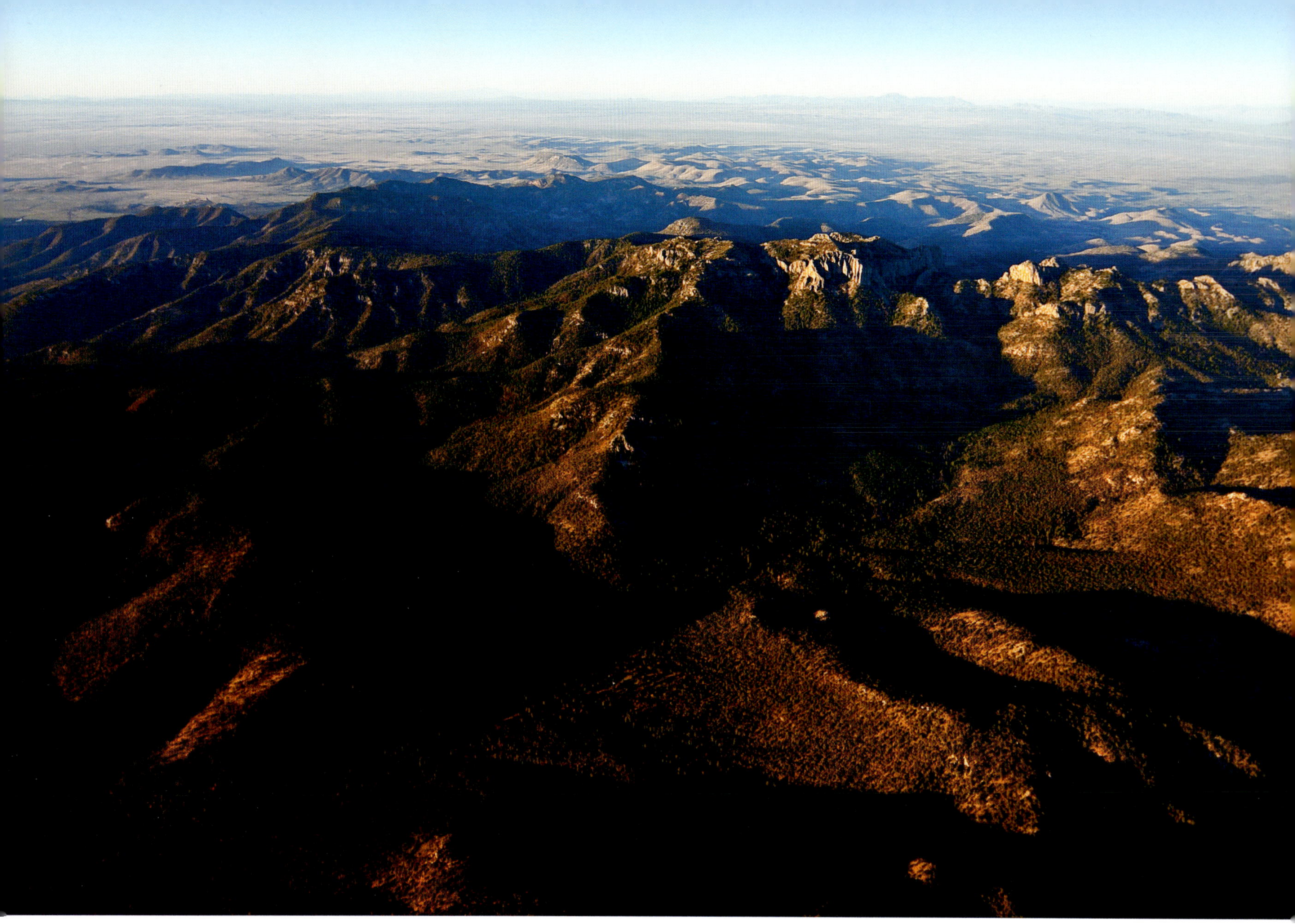

118. Mount Livermore

Davis Mountains

Mount Livermore, with its domed high point known as Baldy Peak, is the highest mountain in the Davis Mountains at 8,378 feet elevation. When one looks southwest toward the desert from which they rise, the Davis Mountains manifest themselves as a "sky island" type of mountain range, surrounded by plains. Mount Livermore features several springs on its slopes that, in conjunction with the cooler mountain climate and higher precipitation, support a unique ecosystem. The Davis Mountains provide an almost pristine habitat for many species of wildlife and plants, including a mind-boggling array of birds and even aspen trees. The nights are cool, even in the summer, and regularly freezing in the winter. I remember driving through Wild Rose Pass in a snowstorm to visit Larry Francell in Fort Davis, where he is county commissioner.

119. Turning toward Home

Fourteen miles southwest of Fort Davis

Leaving the Davis Mountains (upper left) behind, we climb to 12,000 feet for a moment. The thin air and desert light fill me with wonder as we return from our final flight. Here the land opens up from dense mountains to the high desert. Far below us are (center, top to bottom) Blue Mountain (elevation 7,286 feet), Texas Highway 166, North Fork Alamito Creek, and Big Hill. I felt compelled to include this photograph for the way it communicates the almost hypnotic experience of flying at high altitude without pressurized oxygen. The surreal wide-angle view captures the zenith of our flight and the rapture of the heights.

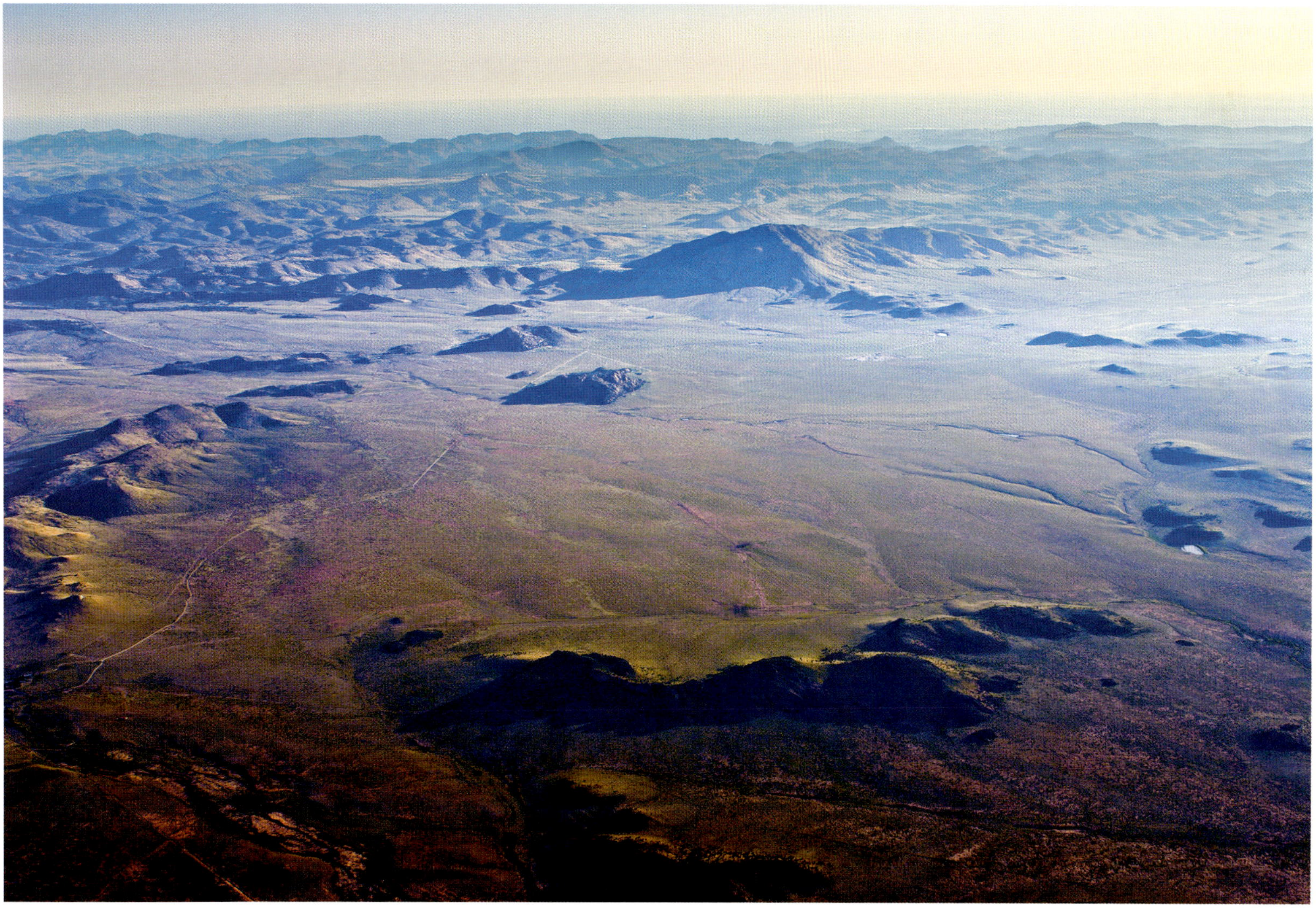

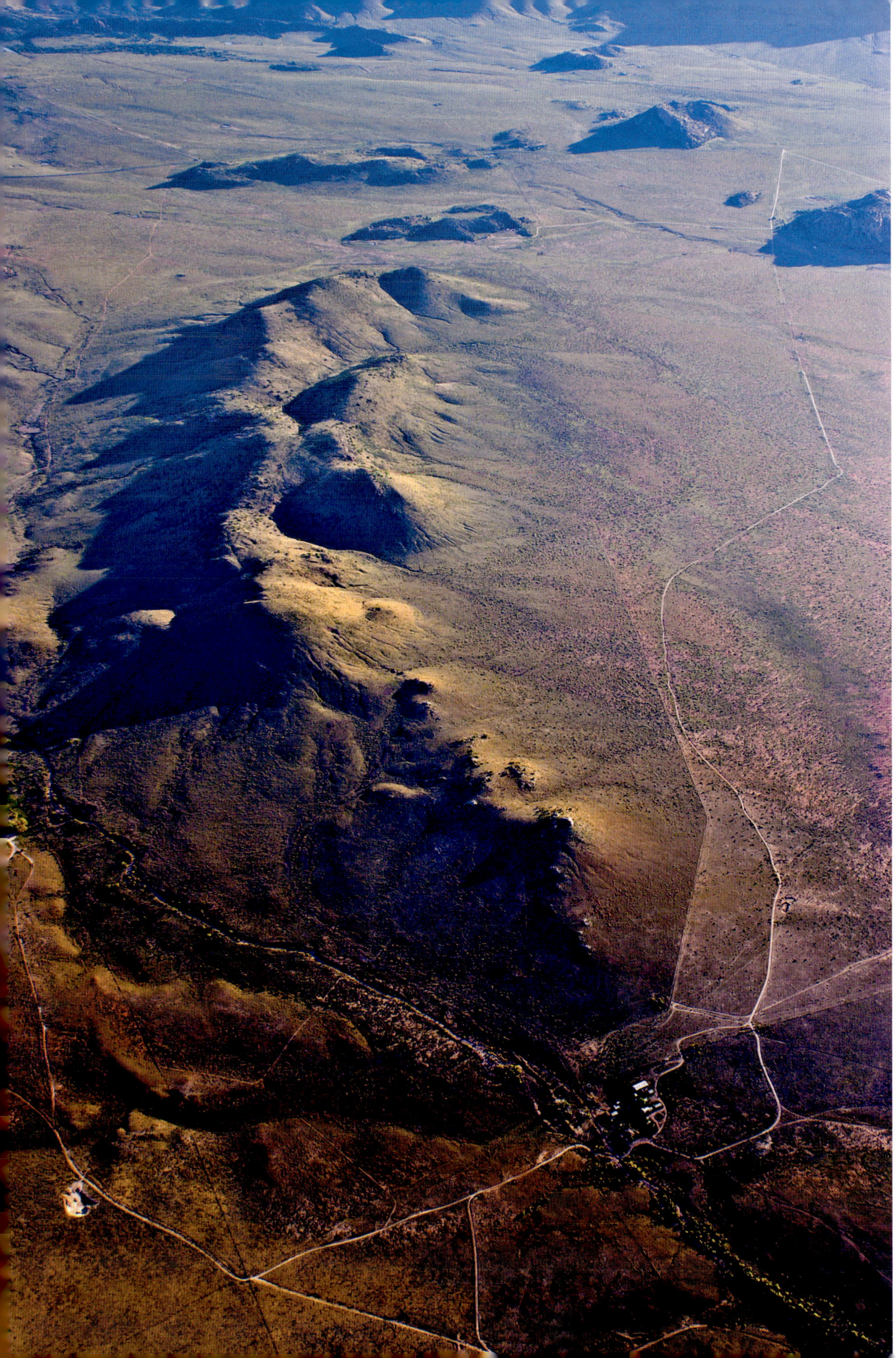

120. Chalk Hill

Ten miles northwest of the Marfa airport

With the nose of our Cessna pointing toward Marfa, I feel euphoria from the high-altitude oxygen deprivation as we climb to the edge of the aircraft's limits to get a final glimpse of this land from above before descending back to earth. Here the air is cold, but below us the desert has heated and dust devils are starting to swirl. Seeing the ranch house just below Chalk Hill in the foreground reminds me that my sojourn in Marfa is coming to an end.

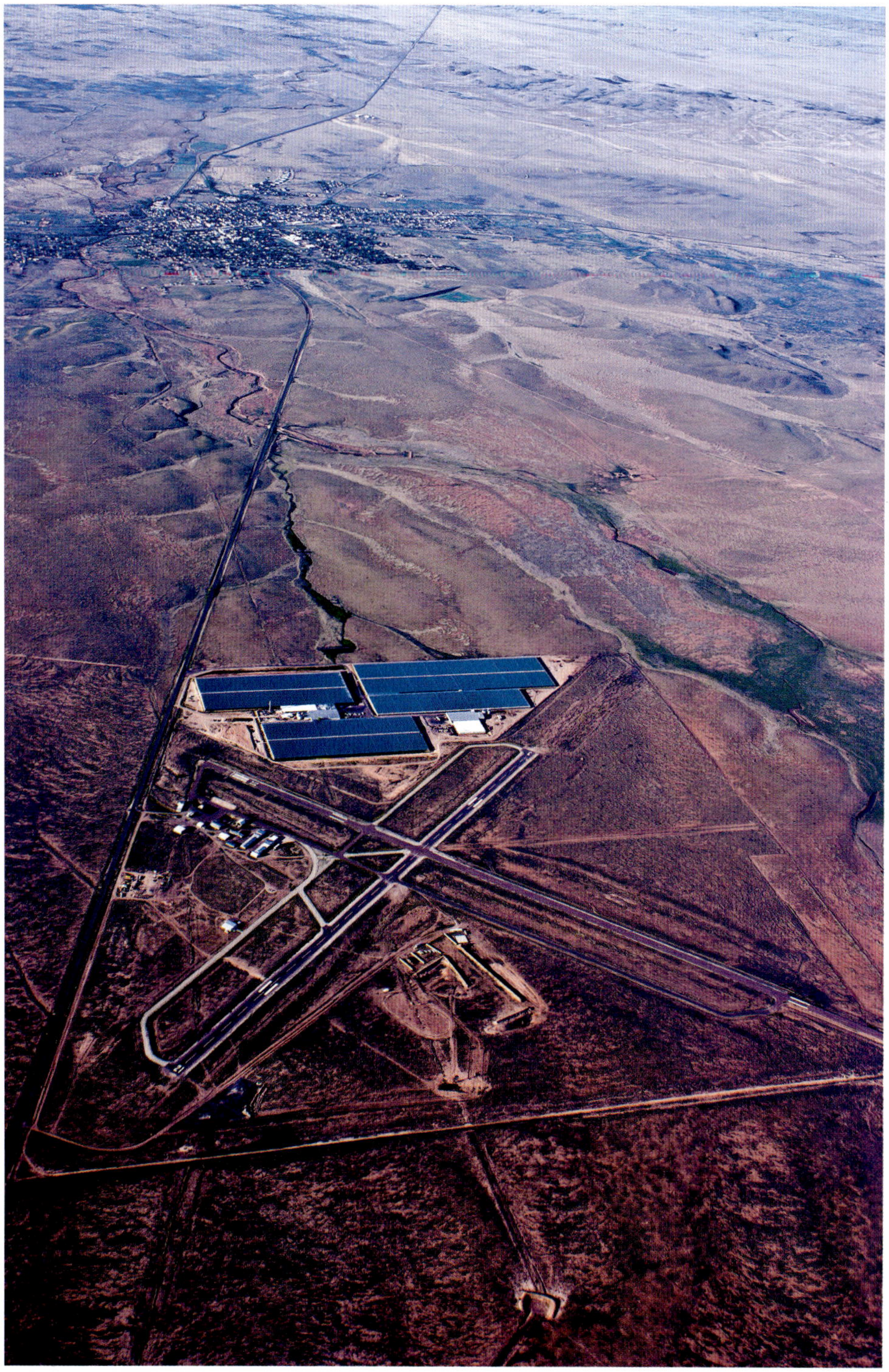

121. The Return to Marfa

Marfa Municipal Airport

In the foreground the two main runways of the Marfa Municipal Airport form a cross in the desert. This airport served as our base for the aerial photography in this book. The large buildings are hydroponic greenhouses that grow tomatoes, cucumbers, and peppers using a sustainable approach. In the distance, four miles to the south, is the town of Marfa (top, left). A vivacious art colony has sprouted around the Chinati Foundation, which was founded by sculptor Donald Judd, like desert wild flowers after a spring rain.

Epilogue

At the end of my artist's journey, we land back in Marfa after the final flight. I would never have imagined that this little town, population 1,966, would represent my return to civilization and my day-to-day life. Back on the ground, my guide and pilot Roger wishes me safe travels home.

Driving north up Texas Highway 17, I see that the Indian paintbrushes and bluebonnets are in bloom. Dallas is an eight-hour trip ahead of me. As I drive, I can still feel the motion of the aircraft and the sensation of seeing the desert, the Chisos Mountains, and the Rio Grande from the air for the first time. In my lingering bliss, I realize the depth of what I have experienced on this adventure that started with a passing thought. Such is inspiration.

On this path that is so mesmerizing and fulfilling, I have returned to my roots. Now I continue this journey by sharing what I have brought back from my sojourn to Marfa. Making this book and exhibiting the photographs in museums and galleries continue to renew my passion and spark ideas for the next project.

If there is a lesson here, it is to do what you love, however imperfectly—and with whatever you have on hand. Then, as Thoreau said, you will "meet with success unimagined in common hours."

Index

Page numbers in *italics* indicate pictorial content in photographs and map.
Page numbers in **boldface** indicate definitions.

Other titles in the Tarleton State University Southwestern Studies in Humanities series

In the Deep Heart's Core: Reflections on Life, Letters, and Texas
Craig E. Clifford

Cannibals and Condos: Texans and Texas along the Gulf Coast
Robert L. Maril

Rough and Rowdy Ways: The Life and Hard Times of Edward Anderson
Patrick Bennett

Larry McMurtry and the Victorian Novel
Roger W. Jones

Nueva Granada: Paul Horgan and the Southwest
Robert F. Gish

Beautiful Swift Fox: Erna Fergusson and the Modern Southwest
Robert F. Gish

Texas Women Writers: A Tradition of Their Own
Grider and Rodenberger

Riding the Wind and Other Tales
James Hoggard

State of Mind: Texas Literature and Culture
Tom Pilkington

Culture in the American Southwest: The Earth, the Sky, the People
Keith L. Bryant

Donald Barthelme: The Genesis of a Cool Sound
Helen M. Barthelme

West of the American Dream: An Encounter with Texas
Paul Christensen

San Antonio on Parade: Six Historic Festivals
Judith B. Sobré

Let's Hear It: Stories by Texas Women Writers
Grider and Rodenberger

Lone Star Chapters: The Story of Texas Literary Clubs
Betty H. Wiesepape

The Pastoral Vision of Cormac McCarthy
Georg Guillemin

Exploding the Western: Myths of Empire on the Postmodern Frontier
Sara L. Spurgeon

Undaunted: A Norwegian Woman in Frontier Texas
Charles H. Russell

The Lost Years of William S. Burroughs: Beats in South Texas
Robert E. Johnson

The Birth of a Texas Ghost Town: Thurber, 1886–1933
Gentry and Baker

Alexandre Hogue: An American Visionary—Paintings and Works on Paper
Susie Kalil

Fritos® Pie: Stories, Recipes, and More
Kaleta Doolin

Faded Glory: A Century of Forgotten Texas Military Sites, Then and Now
Alexander and Utley

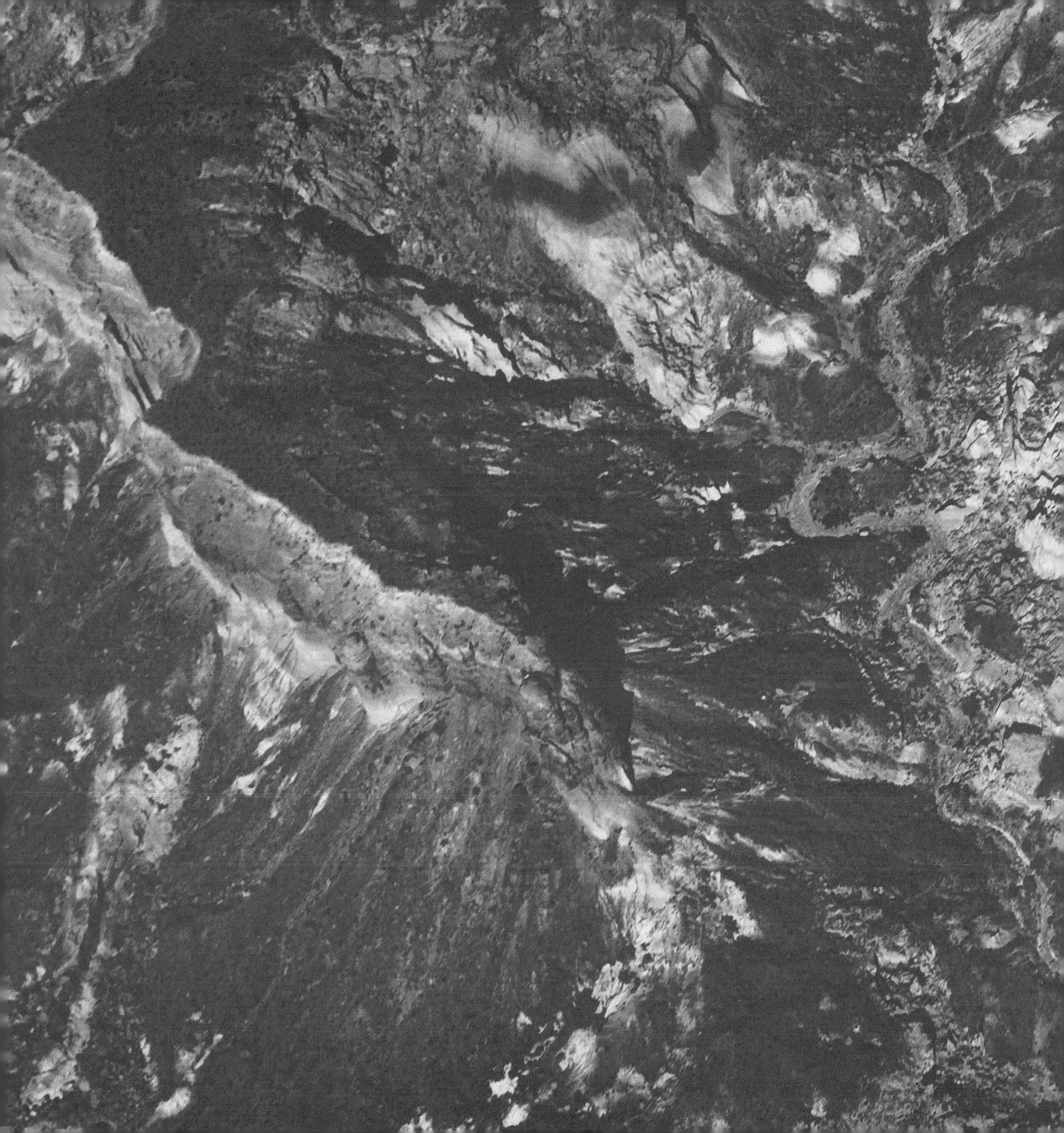